Heinrich Franz Alexander Eggers

The Flora of St. Croix and the Virgin Islands

Heinrich Franz Alexander Eggers

The Flora of St. Croix and the Virgin Islands

ISBN/EAN: 9783743320437

Manufactured in Europe, USA, Canada, Australia, Japa

Cover: Foto ©berggeist007 / pixelio.de

Manufactured and distributed by brebook publishing software (www.brebook.com)

Heinrich Franz Alexander Eggers

The Flora of St. Croix and the Virgin Islands

ST. CROIX AND THE VIRGIN ISLANDS,

BY

BARON H. F. A. EGGERS.

———◆———

WASHINGTON:
GOVERNMENT PRINTING OFFICE.
1879.

ADVERTISEMENT.

This work is the thirteenth of a series of papers intended to illustrate the collections of natural history and ethnology belonging to the United States, and constituting the National Museum, of which the Smithsonian Institution was placed in charge by the act of Congress of August 10, 1846.

It has been prepared at the request of the Institution, and printed by authority of the honorable Secretary of the Interior.

SPENCER F. BAIRD,

Secretary of the Smithsonian Institution.

SMITHSONIAN INSTITUTION,
  Washington, May, 1879.

FLORA OF ST. CROIX AND THE VIRGIN ISLANDS, WEST INDIES.

BY BARON H. F. A. EGGERS.

To the east of the island of Porto Rico, between 18° 5′ and 18° 45′ N. lat. and 64° 5′ and 65° 35′ W. long., stretches a dense cluster of some larger and numerous smaller islands for a distance of about 85 miles, which are known by the name of the Virgin Islands. The principal islands are Vieques and Culebra, belonging to Spain, St. Thomas and St. Jan, belonging to Denmark, and Tortola, Virgin Gorda, and Anegada, belonging to England. The superficial area of the larger islands is only from 16 to 40 square miles, whilst the smaller ones are mostly uninhabited islets, or even rocks, some of which are nearly devoid of vegetation, the coast-line of them all being sinuous, and forming numerous small bays and creeks. The whole group is evidently a submarine prolongation of the mountains of Porto Rico, showing its tops and higher ridges above the level of the sea, the depth of which between the various islands and Porto Rico is only from 6 to 20 fathoms. The declivities to the north and the south of the ridge on the reverse are very steep, no bottom having been found 25 miles to the south in 2000 fathoms, and 80 miles to the north the Challenger Expedition found a depth of about 3850 fathoms, the greatest ever measured in the northern Atlantic Ocean.

The greatest height in the Archipelago is attained in its central part, St. Thomas reaching up to 1550′, Tortola even to 1780′, St. Jan and Virgin Gorda being a little lower, whilst the hills in Vieques and Culebra, to the west, are only 500′–600′ high, and Anegada, the northeasternmost, is, as its Spanish name, the inundated, implies, merely a low or half-submerged island, elevated but a few feet over the level of the sea. The central islands, therefore, present the appearance of a steep ridge, precipitously sloping to the north and the south, and cut up by numerous ravines, which during heavy rains are the beds of small torrents, but which generally are without running water, and which at their lower end widen into small level tracts on the sea-coast, often forming a lagoon on the sandy shore. Between these level tracts the coast is usually very bold and rocky, forming abrupt promontories of considerable height and picturesque appearance, the hills and ridges on the other hand being more rounded and of a softer outline.

The whole group of islands, with the exception of Anegada, which is built up of a tertiary limestone of very recent and probably pliocene date, belongs to the cretaceous period,*showing as the principal rock a breccia of felsite and scoriaceous stones, the cementing part of which probably consists of decomposed hornblende, and having its cavities commonly filled with quartz or calcareous spar. Besides this principal rock, which is often found distinctly stratified, and which is called Bluebit by the inhabitants, who generally employ the stone for building materials, limestone, diorite, clay-slate, and other less frequent minerals also occur in the islands, forming, however, only a poor substratum for vegetation everywhere. For the product of the decomposed rock is generally a red heavy clay. Only Vieques shows a more fertile soil, produced by the alteration of a syenite-like diorite, its more level surface at the same time allowing the fertile strata to remain on the surface; whilst in the other islands the heavy rains as a rule will wash the loose covering of the ground down to the sea.

From various facts observed in Anegada and Virgin Gorda by Sir R. Schomburgk,† as well as by Mr. Scott, in Vieques, at Porto Ferro Bay, it appears that at the present period the whole chain of islands is slowly rising, so that perhaps in a geologically speaking not very distant time most of the islands may become connected reciprocally and with Porto Rico.

To the south of the Virgin Islands, at a distance of about 32 miles, and between 17° 40′ and 17° 47′ N. lat., 64° 35′ and 64° 54′ W. long., lies the island of St. Croix, geographically considered an outlying part of the former group, but separated from it by an immense chasm of more than 2000 fathoms, as stated above. This extraordinary crevice has no doubt been formed at an early period, and has in various respects contributed materially to isolating the island from its neighbours.

St. Croix is of about 57 square miles, and has a triangular form, with the greatest length, some 20 miles, from east to west, the greatest breadth being about 5 miles, in the western part of the island, which becomes gradually narrower towards the east. The coast-line is more connected and the surface more level than in most of the Virgin Islands, the hills stretching only along the northern coast and through the eastern part of

---

*Cleve: On the Geology of the North-eastern West India Islands. Stockholm, 1871.
†Berghaus: Almanach für das Jahr 1857, pp. 405 and 408.

the island, reaching in some places as high as 1150' (Mount Eagle), but averaging 600'–800' only.

The rock of these hills is nearly the same as in the above-named group, although the Bluebit of this latter occurs more rarely, and is substituted by a fine, greyish, stratified clay-slate, without vestiges of any organic remains. The strata of this slate are often very much disturbed, so as to present an exceedingly broken and overturned appearance. The greater, western part of the island forms a large, slightly inclined plain, sloping towards the south, and interrupted in a few places by low, short, isolated ridges only 200'–300' high, and formed of a tertiary limestone of the miocene period. This limestone is covered by a layer of detritus and marls some feet thick, but shows itself at the surface in various places, and contains several fossils, partly of still existing species of mollusca.

Along the coasts are found some new alluvial formations, often enclosing lagoons, some of which are of considerable size. These lagoons are being gradually filled up by vegetable matter, as well as by sand and stones washed down by the rains from the hills; but whilst in the Virgin Islands many similar lagoons have been raised already several feet above the level of the sea, and laid completely dry, no such thing has been observed in St. Croix. This seems to indicate that no rising of the ground is taking place in the latter, as is the case in the former, as mentioned above. From its whole structure and formation it may be inferred that the soil is more fertile in St. Croix than in most of the Virgin Islands, Vieques excepted, the sugar-cane being cultivated to a considerable extent on the island.

Whilst thus the geology of St. Croix and the Virgin Islands presents some not unimportant differences, the climate may, on account of their similar geographical position, as well as elevation above the sea-level, be said to be materially the same in both.

In accordance with the geographical position of the islands, the temperature is very constant and high, the yearly mean average being 27.2° C., divided nearly equally over all the months, the coldest, February, showing 25.6°, the warmest, September, 28.9°, a difference of 3.3° only. The same uniformity is observed in the daily variation, which scarcely ever surpasses 5°, the thermometer rising gradually from 6 a. m. till 2 p. m., and falling just as gradually during the rest of the 24 hours.

Thus the difference of temperature at the various seasons of the year is too small to affect the life of vegetation to any very perceptible ex tent, and it is therefore the variable degree of moisture at different times which chiefly produces any variation in the development of vegetable life at the different seasons.

The lowest temperature observed at the sea-level, in the shade, is 18.1°; the highest, 35.5°. In the sun, the mercury will sometimes rise as high as 51°, but as a rule does not surpass 40°. Observations made in St. Thomas by Knox * and myself show a decrease of about 2° for an elevation of every 800', which gives to the highest ridges in St. Thomas and Tortola an annual mean temperature 3½°–4° lower than that of the coast, a difference sufficient to produce some variation in the flora of these parts. The northern slope of the hills, from being the greater part of the year, viz, from August to May, less exposed to the rays of the sun, are generally also somewhat cooler and more moist than the southern ones, the consequences whereof are also felt in the life of plants to a considerable extent.

An equal regularity, as observed in the temperature, manifests itself with regard to the pressure of the atmosphere, the daily variations of the barometer being only about 0.05'', and the maximum yearly difference only 0.2''. It is only during strong gales and hurricanes that the barometer is more seriously affected, it then falling sometimes as much as 2''. These hurricanes, as a rule, occur only during the months from August to October, at which period the trade-winds from the northeast, which otherwise blow most part of the year, generally become unsteady and uncertain. These constant winds, combined with the high temperature, no doubt are the reason why the moisture of the air is comparatively small, being on an average only 73 per cent. of the possible maximum, thus exciting a constant evaporation in plants, and rendering it necessary for them to obtain a greater supply of water through the soil than in more moist climates. For this reason a considerable quantity of rain becomes of the highest importance to the vegetable life, as being the only form in which plants can obtain a sufficient amount of water necessary to their existence, even dew being very rare on account of the trade-winds blowing also during the night the greater part of the year.

Neither of the islands in question is of sufficient elevation above the sea to cool and condense the atmospheric moisture brought on by the trade-wind, nor is their configuration favourable for detaining the clouds, their greatest extent being parallel to the direction of the wind. Thus, for the greater part of the year they receive only a small quantity of

---

* Knox : An Historical Account of St. Thomas, W. I. (New York, 1872.)

rain, falling chiefly in the form of short, rapid showers of only a few minutes' duration, and it is not till the warmer part of the year that heavy and general rains become possible in these regions. During this latter time, the trade-winds become irregular and slight, or are even entirely suspended, as stated before; hence the moisture generated by the daily evaporation from the ocean is not carried off as soon as formed, but is allowed to gather into rain-clouds, and finally to precipitate itself again as rain nearly on the same spot where it was formed.

From observations made in various islands for a period of more than twenty-five years, the annual mean quantity of rain seems to be about the same in all the islands, averaging 42″–44″; the eastern parts of all, as being more exposed to the direct action of the winds, always showing a considerably smaller quantity than the central and western ones.

Although no month of the year is without rain, yet from the above it will be easily concluded that there is a remarkable difference between the various months in this respect: the driest, February, having only an average of 1.5″; the wettest, October, of 7.0″; and to this difference, at the various periods of the year, it is chiefly due, that notwithstanding the uniform temperature all the year round, yet some variations in the aspect and intensity of vegetable life are observed in the various seasons.

Both the annual and the monthly quantity of rain are subject to vary considerably, one year showing 23″, or in some places 18″ only, another again 70″ or 78″. A still greater difference may be observed between the same months of different years: thus, February having had one year 0.19″ only, another, on the contrary, 3.75″; May 0.47″ the one year and 16.84″ the other. These excessive variations must, no doubt, materially affect vegetable life, indicating at the same time a considerable degree of hardiness in respect to drought in the perennial plants indigenous to the islands, and as alluded to above, acting upon them in a similar way as the variations in temperature in colder climates.

The number of days on which rain falls averages for the period from 1852–73, 161 a year, giving a mean fall of rain of 0.27″ per diem: April showing the lowest number, 9; October the highest, 16. From what has been said before, it is evident, however, that the small monthly quantity of rain during the dry part of the year, viz, January to April, divided even over a great number of days (so as to amount to 0.14″ or 0.18″ only a day), can be of no great importance, as it is precipitated in a short shower, is insufficient for penetrating into the soil, and so is very soon evaporated again by the action of the sun and the trade-wind combined. It is not till May, when the increased quantity of rain is sufficient to penetrate the parched soil, that its influence and effect upon vegetation makes itself felt by renewed life and activity in all the various branches of the vegetable kingdom in general.

Looking at the vegetation of St. Croix and the Virgin Islands in its generality, and without entering into details, we may consider it to be identical, as a whole, showing the same main features, and naturally divided into four distinct formations, as in most other West India Islands, viz. the littoral, the shrubby, the sylvan, and the region of cultivation, connected, of course, here and there by intermediate formations, but on the whole virtually distinct from different biological conditions.

Beginning with the littoral flora, we find along the coast in shallow water a multitude of Algæ, among which are found some marine Phanerogamæ, especially the common *Thalassia testudinum* and *Cymodocea manatorum*, and in less quantity the beautiful little *Halophila Baillonii*, a recently discovered Potamea, with oval delicate leaves, and growing gregariously on the bottom of the sea in coarse gravel. The vegetation of tropical seashores is of a very uniform character all over the world, the physical conditions being similar on them all, and the migration from one shore to another being exceedingly facilitated by the sea as well as by birds, storms, and the action and intercourse of the inhabitants. Thus, the same species of littoral plants are found on nearly all the West India islands, many of them also inhabitants of far distant shores on the African and Asiatic continents,—belonging to the cosmopolitan and transoceanic species, a list of which was first prepared by Robert Brown, and afterwards augmented by A. DeCandolle, and which seem to possess an extraordinary faculty for migration. According to the different character of the coast, as sandy, rocky, or swampy, the vegetation on it also assumes a different aspect.

On the sandy shore, which is composed of a fine white gravel, consisting principally of innumerable pieces of broken shells and corals, and thus forming a thick layer of carbonate of lime, we see a luxurious flora of trees, shrubs, and minor plants, which all, on account of the underground water collecting from the hills above, generally have a green appearance all the year round, even when the hills of the interior present a withered aspect from want of rain. Among the trees growing here the most prominent are the *Hippomane Mancinella*, the *Cocco-*

loba urifera, *Chrysobalanus Icaco*, and *Canella alba*, besides the *Cocos nucifera*, which is planted and naturalized, especially on the low sandy seashore. Under these taller forms appear many kinds of shrubs, such as *Ecastophyllum Brownei*, *Tournefortia gnaphalodes*, *Borrichia arborescens*, *Ernodea litoralis*, *Suriana maritima*, *Erithalis fruticosa*, *Colubrina ferruginosa*, *Guilandina Bonduc* and *Bonducella*, and several others. Still lower shrubs and suffrutescent herbs are *Scævola Plumieri*, *Tournefortia gnaphalodes*, *Sesuvium portulacastrum*, *Heliotropium curassavicum*, *Philoxerus vermiculatus*, *Cakile æqualis*, as well as several grasses and sedges, as *Sporobulus litoralis*, *Stenotaphrum americanum*, and *Cyperus brunneus*, as also some remarkable creepers or climbers, such as *Ipomæa pes-capræ* and *Lablab vulgaris*.

Most of these species disappear on the rocky cliffs, where they give room for others, mostly shrubs of a low growth, and with thicker or more coriaceous leaves, that are able to resist the force of the wind, which often bends the whole plant into a dwarfish individual, the branches of which are cut off at the top in a western direction. The most common of these shrubs are *Jacquinia armillaris*, *Elæodendron xylocarpum*, *Plumieria alba*, and *Coccoloba punctata*, as well as some monocotyledonous plants, such as *Pitcairnia angustifolia*, *Agave americana*, and a few Cacti, principally the stout *Melocactus communis*.

Still more different forms appear where the coast becomes swampy from the presence of lagoons. Here predominates the Mangrove formation, composed chiefly of *Laguncularia racemosa*, *Conocarpus erectus*, *Avicennia nitida*, and *Rhizophora Mangle*, which all grow more or less in the water itself. In less moist places we find some others, such as *Bucida Buceras*, *Anona palustris*, *Antherylium Rohrii*, and the curious *Batis maritima*, which recalls to the mind the halophytes of the steppes.

However different these various forms of littoral plants may appear, compared to each other, yet they all have in common the predilection for the sea, the saline exhalation of which seems indispensable to their growth. Some have even, like *Avicennia*, their leaves always covered with small salt crystals; others, like *Batis maritima*, are true halophytes, and only very few of the plants of the coast in generality are found in the interior even of these small islands. An exception is made by the cocoanut palm, which is found growing all about on the islands, even on the top of the highest hills, as also by *Coccoloba urifera*, found in similar localities.

In passing from the coast into the interior we find on the eastern, and

partly also on the southern part of all the islands, a dry shrubby vegetation of a greyish or yellowish aspect, which, from the predominating
genus composing its elements, I have called the Croton vegetation. This
peculiar kind of dry shrub also occurs here and there in other parts of
the islands, where the soil, through reckless cultivation, has become too
exhausted to produce a growth of taller trees, and it cannot be estimated
to cover less than one third part of the whole surface of the islands, predominating in some, as Tortola, St. Thomas, and Culebra, less conspicu-
ous in others, as St. Jan, Vieques, and St. Croix.

The ravines as well as the northern and western parts of the islands
are often covered with a growth of taller trees, forming a kind of forest,
composed of species partly evergreen and partly with deciduous foliage,
and which, from one of the most prominent forms, I have called the Eriodendron vegetation. The area covered by this formation may be taken
to be about one fifth of the whole surface, the best wooded islands being
St. Jan and Vieques, the least wooded ones St. Thomas and Virgin
Gorda.

The remainder of the surface is either used for pasture or cultivated
with sugar-cane or provisions, the former on a large scale in St. Croix
and Vieques only, the latter everywhere on the islands where the soil
seems proper for the purpose. This last section I term the cultivated
region.

Considering first the Croton vegetation, we find here a number of plants
which in various ways have become enabled to resist the deteriorating
effects of the dry climate, and to exist on the barren rocky soil always
found where the moisture is not sufficient for decomposing the natural
rock of the surface. Thus, some of these plants, as the whole of the
genus Croton, already mentioned above, have small leaves, which, like
the stem, are covered with scales and tomentose hair, containing besides
aromatic oil, all which contrivances tend to diminish evaporation as much
as possible. The most common species of this remarkable genus are *C.
flavus, astroites, bicolor,* and *betulinus.* Other forms obtain the same
object by having very small, partly deciduous leaves and their stipules
transformed into prickles, especially the Acaciæ, such as *A. Farnesiana,
macracantha, tortuosa,* and *sarmentosa.* Others, again, are rich in milky
juice, as *Euphorbia petiolaris, Rauwolfia Lamarckii,* and the naturalized
Calotropis procera, or merely in aqueous sap, as the Cacteæ, the commonest forms of which are *Melocactus communis, Cereus floccosus,* and several
species of Opuntia. Others, such as Bromeliaceæ, on the contrary, have

a very dry structure, and a dense cover of scales for protection, whilst others again, such as *Anona squamosa*, which are apparently without any means to resist the effects of dry weather, have no other remedy left than to shed their leaves during a part of the year, and thus preserve their existence at the temporary sacrifice of their vegetative organs.

All the forms mentioned above are of very slow growth, and, with the exception of a few that are used for burning charcoal, of scarcely any importance either to man or animals, for which reason the districts occupied by them as a rule present a very desolate and uninviting appearance.

Where the climate becomes sufficiently moist, and the soil in consequence thereof more decomposed and fertile, the forest appears in place of the Croton vegetation, on the uncultivated lands, especially in ravines and on steep declivities, which do not allow of cultivation or grassfarming. As nearly everywhere in the tropics, the forest here is composed of many different species of trees mixed together, a gregarious growth being very rare. From the forests of moister tropical countries, however, the woods in these islands are distinguished by possessing a quantity of forms with thin, herbaceous leaves, which for this reason shed their foliage during a part of the year, thus combining the appearance of the woods of colder climates with the dark evergreen forms of the intertropical countries. Some of these species with deciduous foliage have two periods for flowering: one precocious in the first months of the year, when the small quantity of rain seems insufficient to produce both leaves and flowers at a time, and another later in the year, when both foliage and blossoms are vigorously developed by the increased moisture of the summer. The evergreens for the same reason have a less fixed and more unlimited time for flowering, and seem to show their reproductive organs whenever the quantity of rain becomes sufficient for producing them besides maintaining the already existing foliage. Among the great variety of evergreen forms of trees and shrubs, I shall here only mention as the most common several species of Anona; of Guttiferæ, such as *Calophyllum Calaba* and *Clusia rosea;* of Sapotaceæ, such as Sideroxylon, Chrysophyllum, Lucuma, and Dipholis; of Rutaceæ, as Zanthoxylum and Tobinia; of Lauraceæ, as Nectandra and Oreodoxylon, as well as many others, for the details of which I beg to refer to the systematical part of my treatise. Others are possessed of aërial roots by which to affix themselves to the stems of trees and rocks, as several species of Ficus; others again are vines, such as Bignonia, Serjania, Gouania, and Cissus.

Interspersed between these evergreens are seen various species of arboreous plants with deciduous leaves, the number of which, however, seldom is large enough to seriously change the general aspect of the forest as being uniformly green all the year round. The time for shedding their foliage in these forms is generally from January to April, most of them, as stated before, flowering precociously at this time, as the moisture in the ground is not sufficient to allow them to retain their foliage together with the producing of the flowers. It appears evident that this is the reason for the shedding of the leaves, from the fact observed by me in several species (such as *Piscidia Erythrina* and others), that individuals which, from being too young or for some other reason, do not flower, do not shed their foliage, but evidently find moisture enough in the soil to resist the drought, not having to spend their resources on the production of flowers and fruits, as others of their kind.

The most prominent among the trees and shrubs with a deciduous foliage are *Spondias lutea*, *Schmidelia occidentalis*, the enormous *Eriodendron anfractuosum*, *Hura crepitans*, *Cascaria ramiflora*, *Sabinea florida*, and several others, which all more than the evergreens contribute their share to the forming of a layer of leaf-mould under the taller forms. Yet this layer is but scanty in most places, and from the want of it, as well as from the dense shade produced by the evergreen trees and shrubs, the minor forms covering the ground are comparatively scarce, and chiefly confined to some Piperaceæ, Acanthaceæ, and Gramineæ, as well as a few ferns and mosses, among which *Hemionitis palmata*, *Pteris pedata*, and *Asplenium pusillum* are the most common.

A somewhat richer variety is presented by the numerous epiphytes that cover the branches and stems of trees and shrubs, notwithstanding that the bark of the latter, from the uniform temperature, is, as a rule, exceedingly smooth, and but rarely covered with lichens or mosses. Of real parasites only a few are met with, especially *Loranthus emarginatus*, whilst the non-parasitical epiphytes are numerously represented by Bromeliaceæ (principally the genus Tillandsia), Aroideæ (among them the large-leaved *Philodendron giganteum*), and Orchidaceæ (chiefly Epidendrums and Oncidiums), as well as some ferns. Of these latter families, several species are found only on the highest ridges of the islands, at an elevation of over 1300', there forming a formation peculiar to these regions, comprising, among others, some terrestrial Orchids, such as *Habenaria maculosa* and *alata*, as well as some Aroideæ, Bromeliaceæ, and ferns, among which the beautiful *Cyathea arborea* deserves special mention.

The part of the island inhabited and cultivated by man of course represents the least of interest in a phyto-geographical sense, as nature here has been modified and modelled according to the wishes and necessity of society to such an extent as to almost entirely obliterate its original character. As stated already, the principal object of cultivation is the sugar-cane, which, however, is cultivated on a large scale only in the two largest and most level of the islands, Vieques and St. Croix, the others, viz. St. Thomas, St. Jan, Tortola, and Virgin Gorda, having, with a few exceptions, long ago abandoned the cultivation of the cane as unremunerative, the two remaining of the larger islands, Culebra and Anegada, never having been appropriated to that purpose.

Besides the cane, some *Sorghum vulgare* is also cultivated in fields for herbage, the rest of the tilled soil being used for the planting of the common tropical vegetables, generally in small quantities, on patches of soil selected here and there. The commonest of these plants are Yam (*Dioscorea alata* and *altissima*), Sweet Potato (*Ipomœa Batatas*), Okro (*Abelmoschus esculentus*), Tanier (*Xanthosoma sagittæfolium*), Pigeon-pea (*Cytisus Cajan*), Tomato, and Pepper (*Capsicum*), as well as some Cucurbitaceæ, as Pumpkin, Melon, and others.

Along with these useful plants follow a great number of herbaceous annuals, mostly cosmopolitan weeds, introduced after the settlement of the islands, and dependent on the continuous cultivation of the land, as without the clearing of the soil from shrubs and trees their existence would soon be terminated by the stronger arboreous species, which would deprive them of the necessary light and air.

Thus, much against his wish, man favours the propagation of innumerable weeds, which in their short period of vegetation produce seeds enough to secure their continuance on the land notwithstanding the efforts to exterminate them by frequent weeding. Among the commonest of these forms are some Labiatæ (*Leonurus sibiricus, Leonotis nepetæfolia*, and *Leucas martinicensis*), *Argemone mexicana, Tribulus maximus, Boerhaavia erecta* and *paniculata*, and especially many grasses and sedges, such as Panicum, Paspalum, Chloris, Digitaria, Cyperus, and others. The most troublesome of these, from an agricultural point of view, is the Bay-grass (*Cynodon Dactylon*), said to be introduced, but now found everywhere, and, on account of its long creeping rhizoma, inexterminable.

Similar forms to these are seen growing along roads and ditches, especially some Leguminosæ, as Crotalaria, Desmodium, Phaseolus, Clitoria,

Centrosema, Teramnus, Vigna, Rhynchosia, and others; grasses, as Lappago, Aristida, Sporobolus, Eleusine, Dactyloctenium, and Eragrostis; or Synantheræ, as Elephantopus, Distreptus, Bidens, and Pectis. Whilst all these latter forms flower during the greater part of the year, the beautiful Convolvulaceæ, such as *Ipomœa fastigiata, Nil, umbellata, dissecta, violacea*, and others, are in blossom only during the winter months, from December to February.

In some places that are moist enough, sedges and semi-aquatic plants will be seen growing; in a few rivulets which contain water all the year round, and which are limited to Vieques and St. Croix, a few aquatic forms occur, such as *Echinodorus cordifolius, Lemna minor, Typha angustifolia*, and *Nymphæa ampla*.

The pastures, which occupy a considerable extent of the land, are either artificial,—planted with Guinea-grass (*Panicum maximum*), a perennial plant, and, like most of the cultivated West India plants, introduced from the Old World,—or natural, covered with various forms of indigenous Gramineæ as well as low shrubs and trees, that have continually to be cleared away to prevent the land becoming overrun by them. The artificial pastures as a rule are fenced in, and often protected against the dry season by the planting of Thibet-trees (*Acacia Lebbek*), now commonly naturalized everywhere; the natural ones, on the contrary, are generally open and abandoned to the cattle, whilst the artificial ones are cut regularly, and the stock is not allowed to enter them.

The grasses composing the natural pastures are several species of Panicum, Paspalum, Dactyloctenium, and Sporobolus; some, as *Tricholæna insularis*, being very bitter and unfit for herbage. The roaming about of the cattle everywhere effectually prevents the re-establishment of trees or woods, which, for climatic reasons, would be desirable in many places; for the young buds are destroyed by sheep and goats, which no doubt have contributed largely to deteriorating even the still existing woods.

Around dwellings are found planted and naturalized most of the plants now common to nearly all tropical countries,—some fruit-bearing, as *Tamarindus indica, Mangifera indica, Carica Papaya, Persea gratissima, Crescentia Cujete, Melicocca bijuga;* others ornamental, as *Poinciana regia, Calliandra saman, Cæsalpinia pulcherrima*, and others Actual gardens are now very rare, flowers being mostly cultivated in pots or boxes. Some few vegetables of colder climates are cultivated in shady places where water is abundant, such as salad, radishes, cabbage, and others. In waste places are found most of the tropical weeds, as *Ricinus com-*

munis, Datura Metel and *Stramonium, Euphorbia pilulifera, heterophylla,* and *hypericifolia, Mirabilis jalapa, Jatropha curcas, Cassia occidentalis,* and especially several kinds of Sida and Abutilon as well as some other Malvaceæ.

The four formations mentioned above are usually found only on the larger islands, the smaller ones, from their limited size, generally possessing chiefly the littoral and shrubby only. The island of Anegada, although being one of the larger ones, yet from its structure and the nature of its soil, seems to be chiefly covered by a vegetation composed of the plants of the sandy shore, besides some of the trees and shrubs following the settlement of man in these regions. Sir R. Schomburgk, who has given a description of the island in the Journal of the Royal Geographical Society, 1832, asserts that the island possesses several interesting species of plants, among others a peculiar kind of Croton. As, however, I have not been able to procure the work referred to above, I am not prepared to say which those species are, and they are not mentioned by Prof. Grisebach in his Flora of the British West India Islands.

Although, as stated above, the general character of the flora both in St. Croix and the Virgin Islands, considered as a whole, is essentially the same and distinctly West Indian, yet, in looking more closely into details, we are soon struck by finding a great many species in the one which are not found in the other. This is the more remarkable, as from a geographical and climatical point of view the physical conditions must be said to be materially identical.

In referring to the list of plants given at the end of my treatise it will be seen that out of a number of 881 indigenous phanerogamous species no less than 215, or c. $\frac{1}{4}$, are found in the Virgin Islands only, whilst 98, or about $\frac{1}{9}$, occur only in St. Croix, thus leaving only 568, or less than $\frac{2}{3}$, in common to both.

As may be expected from the general character of littoral vegetation, there are very few species which are not found on both sides of the deep channel separating St. Croix from its northern neighbours, the principal exception being *Baccharis dioica,* which only occurs in St. Croix, and *Egletes Domingensis,* found by me only in the Virgin Islands.

Some greater difference is found in the dry shrubby formation, where several very common plants, such as *Euphorbia petiolaris, Acacia sarmentosa, Mamillaria nivosa,* and others, are to be seen in the Virgin Islands only, St. Croix having to itself a few less common species, such as *Securinega acidothamnus* and *Castela erecta.*

It is, however, in the forest vegetation, which best represents the original flora of the islands, that the greatest and most varied differences are observed, showing especially the great variety of species in the Virgin Islands which are not all found in St. Croix, and among which are many of the commonest and most generally distributed forms. Belonging to St. Croix alone are comparatively few and rare species, chiefly some Rhamnaceæ, viz, *Maytenus elæodendroides* and *Zizyphus reticulatus, Catesbæa parviflora, Beloperone nemorosa, Petitia Domingensis, Buxus Vahlii,* and *Urera elata.* All these forms occur only in a few localities, and are of no importance to the general character of vegetation, as is the case on the Virgin Islands with many of the following species that are found on them, but not in St. Croix. It would be too much to mention all the different species here, for which I beg to refer to the appended list and tabular statement. 1 shall only enumerate a few of the most interesting, especially Malpighiaceæ (as *Byrsonima lucida, Malpighia Cnida* and *angustifolia*), Rutaceæ (*Pilocarpus racemosus, Tobinia spinosa, Xanthoxylum ochroxylum*), Leguminosæ (*Sabinea florida, Pictetia aristata, Sesbania sericea,* and *Acacia nudiflora*), and Sapotaceæ (*Sapota Sideroxylon*). Among Monocotyledones are to be mentioned *Arthrostylidium capillifolium, Rhynchospora pusilla, Dioscorea pilosiuscula, Catopsis nutans,* and several Orchids. Several of these plants grow more or less gregariously, thus becoming characteristical to the formation. Among these are *Malpighia Cnida, Reynosia latifolia, Acacia nudiflora, Sabinea florida,* and several species of Pilea, most of them being very common, and even generally used for domestic purposes.

Besides these species, entirely wanting in St. Croix, the Virgin Islands possess several that are very common, or at least not uncommon on them, but which occur but very rarely in St. Croix, such as *Thrinax argentea, Rondeletia pilosa, Faramea odoratissima, Miconia angustifolia, Mimosa Ceratonia,* and others, and most of which I have not found myself in the latter island, but only found labelled with St. Croix as habitat in the Copenhagen herbarium, so that an error in some cases at least may be not at all impossible.

However great are the differences in the flora on the two groups of islands, yet this interesting fact is not due to their possessing endemic species, as all the plants known as growing on them are also found in other West India islands, especially Porto Rico, whence the vegetation of both the Virgin Islands and St. Croix seems to be derived. Thus it is mainly to different periods of immigration under varied physical conditions that we must ascribe the remarkable discrepancies in the flora of those apparently homogeneous islands. Some few species, it is true, are indeed given in my list as having been found only in the Virgin Islands, such as a few Cacteæ, *Vernonia Thomæ*, and the new species described by me on the present occasion. But as long as Porto Rico, Hayti, and even Cuba, are still insufficiently explored, it may very well remain doubtful whether those species do not also occur in one or several of them, just as several Cuban plants, described as endemical in that island by Prof. Grisebach, have been found by me to occur not at all unfrequently in the Virgin Islands and St. Croix, such as *Arthrostylidium capillifolium*, *Reynosia latifolia*, and *R. mucronata*.

It may thus be confidently asserted that both the groups in question have derived their stock of plants from the neighbouring larger island of Porto Rico. The question that remains to be solved is merely why have they not all received the same species, and particularly why is it that St. Croix, although the largest of all, has received a comparatively and absolutely much less number of species than for instance the far smaller St. Thomas?

For the explanation of these interesting facts we have no doubt to look to the geological history of the islands, as the conditions for immigration over sea, even if possible to all the species, are essentially the same in both groups, and therefore give no solution of the problem in question.

I am thus led to think that at a former period all the West India islands have been connected mutually, and perhaps with a part of the American continent also, during which time the plants in common to all the islands, as well as to the West Indies and the continent, have expanded themselves over their present geographical areas, at least as far as they are not possessed of particular faculties for emigration over the sea. By a subsequent volcanic revolution, St. Croix, as well as many of the other islands, has thereafter been separated from Porto Rico and the Virgin Islands, and put into its present isolated position, which it seems to have retained ever since, whilst the latter group of islands has either still for a long period remained in connection with Porto Rico, or, if separated at the same time from it as St. Croix, has, by another revolution, been again connected with the former.

The plants now found in the Virgin Group, but not occurring in St. Croix, would thus have immigrated into the former from Porto Rico after the separation of St. Croix from the latter, and immigration would finally have ceased by the separation between them, as it exists at the present period. Thus, the plants found in the Virgin Islands, but not in St. Croix, would seem to have been more recently created in the probable centre of vegetation, Porto Rico, or some other of the larger Antilles; the endemic ones, as in the other islands also, being the youngest of all, not having been formed till after the complete separation between the islands had been effected. This latter suggestion, which perhaps seems contradictory to the general accepted theory of considering the endemic forms on oceanic isles as the remnants of the oldest original vegetation,* appears to be confirmed by the fact that even on such recent formations as the Bahamas, which have as yet been but imperfectly explored, already no less than eighteen endemic species have been discovered.†

The supposition that the islands may have been separated from the beginning, and have received their floras through immigration over the sea, is sufficiently confuted, partly by the great number of species common to them all, which clearly indicates the connection in former times with a larger country, partly by the circumstance that most of the species common to the islands are in no way better adapted for migration over the water than those peculiar to the Virgin Islands only; in fact, but few of them apparently possess the faculty of crossing salt-water even for a limited distance.

Supposing the theory of a prolonged or oftener repeated connection between Porto Rico and the Virgin Islands to be correct, it remains still to explain how St. Croix can have obtained a number of species which do not occur in the latter group. A few of these species, viz, *Castela erecta*, *Maytenus elæodendroides*, *Zizyphus reticulatus*, *Anthacanthus jamaicensis*, and *Buxus Vahlii*, occur in St. Croix on the tertiary limestone only, and seem thus to have avoided the Virgin Islands as not finding there the substratum suited to their organisation. The greater part, however, might, for all apparent reasons, as well occur in the Virgin group as in St. Croix, and their absence in the former cannot be explained in this way. It must, however, be understood that whilst my investigation of St. Croix has been thorough, and carried on for several years, my exploration of the Virgin Islands has been so for only a part of them, especially the Danish ones, my collections from the

---

* Hooker: On Insular Floras.
† Griseb.: Geogr. Verbr. der Pfl. Westindiens, p. 55.

others being only imperfect. Without expecting too much from this circumstance, yet I feel confident that not few of the St. Croix plants, apparently wanting in the Virgin group, may, by closer research, still be discovered growing there on some of them, whilst, on the other hand, I am equally confident that none, or scarcely any, of the Virgin Islands' species wanting in St. Croix will be found in the latter island.

It may furthermore be observed that scarcely any of the St. Croix species which I have given as being absent from the Virgin group are common or widely distributed over the island, and so are not possessed of any great faculty for conquering ground in the struggle for existence, for which reason some of them may not have been able to gain admission on the much smaller surface of the Virgin Islands, or, having obtained a footing, they may have lost it again by the later immigration of other species, now peculiar to the group compared with St. Croix, many of which, as will be remembered, are gregarious, and gifted with great facility for expanding themselves.

A very few species form an exception as to the limited distribution in St. Croix, *Bacharis Vahlii, Cordia alba,* and *Ægiphila martinicensis,* occurring rather frequently in the island, but having as yet not been found at all in the Virgin group, although they occur in several others of the West India islands. I am not prepared to give a satisfactory explanation of this fact at the present moment; but such isolated exceptions will no doubt always be met with in the explanation of general phenomena, and most probably a more thorough investigation of vegetable biology will at a future day afford a satisfactory explanation of such apparently inconsistent facts.

In drawing the necessary consequences of the above stated theory for explaining the geographical distribution of vegetable species in St. Croix and the Virgin Islands, it would thus appear necessary to conclude, for instance, from the occurrence of *Sabinea florida* both in Porto Rico, the Virgin Islands, and Dominica, but not in St. Croix, that the first-named islands were still all connected, when the latter had already been separated from them and put into its present isolated position. A similar inference might be drawn from the distribution of *Malpighia Cnida,* whilst the occurrence of *Acacia nudiflora* would seem to prove a similar thing for Hayti, Porto Rico, and Antigua.

It can, therefore, scarcely be presumed, as done by Prof. Grisebach in his Geogr. Verbreitung der Pfl. Westindiens, that the distribution of species is regulated chiefly by geographical distances. A closer investigation of the flora of the various islands no doubt will confirm the theory drawn from the facts observed in regard to the mutual relation between St. Croix and the Virgin Islands, that geological revolutions have been equally or perhaps even more powerfully influential in arranging the distribution of species than the greater or smaller distance, and the similarity of physical conditions.

A full knowledge of these interesting facts can, however, not be expected till a more thorough exploration of all the West India islands has taken place. Few of them are as yet tolerably well known, and it is therefore earnestly to be hoped that such an exploration of all the West Indies may soon be effected, the result of which will no doubt be of the highest importance both to botany and to all other branches of natural science.

It generally requires the accumulated study and knowledge of generations before the less palpable and more delicate, but often most important, facts in natural history can be explained: the West Indies have been comparatively well studied since the middle of the last century; and it would seem well now to follow up the work in order to complete a thorough investigation, which might be used as a basis for the explanation of similar facts observed in other and less well known parts of the world.

The flora of the Virgin Islands and St. Croix has been studied by several botanists, some of whom have published the results of their research, which has, however, among the former group, been chiefly confined to the Danish islands, the English and particularly the Spanish ones having as yet been only imperfectly explored.

Publications on the flora of these islands are given by West in his Description of St. Croix (Copenhagen, 1793); Schlechtendal, Florula Ins. St. Thomæ, in Linnæa, 1828–31 and 1834; and Eggers, Flora of St. Croix, in the Vidensk. Medd. fra Naturhist. Forening (Copenhagen, 1876) besides minor contributions in Vahl's Eclogæ Americanæ, Symbolæ Botanicæ, and Enumeratio Plantarum, Krebs in Naturh. Tidsskrift, 1847, on the flora of St. Thomas, De Candolle's Prodromus, and Grisebach's Flora of the British West India Islands. This latter work, no doubt from want of material, scarcely ever mentions the British Virgin Islands.

Collections of plants from the islands in question are found chiefly in the Museum of the Botanical Garden in Copenhagen, as well as scattered in other European herbaria, collected principally by v. Rohr, West, Dr. Ryan, Ledru, Riedlé, L'Herminier in the past century, by Benzon, Wahlmann, Ehrenberg, Dr. Ravn, Dr. Hornbeck, Duchassaing, Schomburgk, Plée, Wydler, Örsted, Krebs, and Eggers in the present.

The following list of plants from St. Croix and the Virgin Islands formed on my own collections and the publications or collections of other botanists, comprises 1013* species of phanerogamous and vascular cryptogamous plants, of which 881 are indigenous and 132 naturalized, those merely cultivated being added in brackets after each family.

In determining the species I have, besides consulting the more important general systematical works on botany, as much as possible followed Prof. Grisebach's standard work on the Flora of the British West India Islands, to which I therefore beg to refer when no other authority is given. Synonymes and references to other authors are given only where it was thought desirable to supplement the Flora of Grisebach in this respect.

To the specific names of plants I have added only such statements as are not given in Grisebach's work,—as local name, time for flowering, technical use, as well as descriptive remarks, where my own observation shows a difference from the description given in the flora mentioned above.

In referring to Schlechtendal, or the herbarium of the Copenhagen Museum, I have used the abbreviations Schl. and Hb. Havn.; in quoting West or Schlechtendal, their respective works on St. Croix and St. Thomas, mentioned above, are understood to be referred to.

Special localities for habitats are given only where a plant is rare, or at all events uncommon; otherwise the island alone is mentioned.

The expression, "All islands," is meant to imply that the species is found both in St. Croix and the Virgin group, without necessarily meaning to say that it occurs in every island of the latter.

In summing up the statistical results from my list of species, nearly the same conclusions with regard to the most numerous families are arrived at as those given in Prof. Grisebach's Geogr. Verbr. der Pflanzen Westindiens, p. 73, for the Caribbean Islands.

The proportion between Mono- and Dicotyledonous plants indigenous and naturalized is 1:5.8, in the indigenous ones alone 1:4.9, thus show: ing the plurality of the recently introduced plants to have been Dicotyledonous. The proportion mentioned in the plants indigenous to the islands is somewhat lower than stated by Grisebach, as cited above, to be the rule in the West Indies, where it is given as 1:4, indicating, no doubt, that the climate of St. Croix and the Virgin Islands is less moist than that of the West Indies in general.

---

* De Candolle (Geogr. Bot. p. 1274) gives to St. Thomas as the probable number of Phanerogama only 450; but my list shows about 900.

Table showing the distribution of the Indigenous Species of Phanerogamæ and Cryptogomæ Vasculares in St. Croix and the Virgin Islands.

	St. Croix only.	Virgin Islands only.	Common to both.	Total.
A.—DICOTYLEDONES.				
Dilleniaceæ	....	1	....	1
Auonaceæ	1	1	5	7
Menispermaceæ	....	1	1	2
Nymphæaceæ	....	....	1	1
Papaveraceæ	....	....	1	1
Cruciferæ	....	....	3	3
Capparidaceæ	....	1	7	8
Bixaceæ	....	1	5	6
Violaceæ	....	....	1	1
Polygalaceæ	....	3	....	3
Caryophyllaceæ	1	2	9	12
Malvaceæ	4	6	21	31
Bombaceæ	1	....	2	3
Büttneriaceæ	1	....	5	6
Tiliaceæ	1	....	7	8
Ternströmiaceæ	....	....	1	1
Guttiferæ	....	....	3	3
Canellaceæ	....	....	1	1
Erythroxylaceæ	....	....	1	1
Malpighiaceæ	....	3	7	10
Sapindaceæ	2	1	4	7
Meliaceæ	....	....	3	3
Oxalidaceæ	....	....	1	1
Zygophyllaceæ	1	....	2	3
Rutaceæ	3	3	3	9
Olacaceæ	....	....	1	1
Ampelideæ	....	....	4	4
Celastraceæ	1	....	5	6
Rhamnaceæ	2	1	4	7
Terebinthaceæ	1	1	5	7
Leguminosæ	7	18	50	75
Chrysobalanaceæ	....	....	1	1
Myrtaceæ	4	4	18	26
Melastomaceæ	....	4	6	10
Lythrariaceæ	....	....	2	2
Onagraceæ	....	....	1	1
Rhizophoraceæ	....	....	1	1
Combretaceæ	....	....	3	3
Cucurbitaceæ	1	1	7	9
Papayaceæ	....	....	1	1
Passifloraceæ	2	1	5	8
Turneraceæ	....	1	1	2
Cactaceæ	....	4	8	12
Araliaceæ	....	1	....	1
Umbelliferæ	....	1	....	1
Loranthaceæ	1	....	1	2
Rubiaceæ	4	8	22	34
Synantheræ	4	13	32	49
Lobeliaceæ	....	....	1	1
Goodenoviaceæ	....	....	1	1
Myrsinaceæ	....	....	2	2
Sapotaceæ	....	2	9	11

Table showing the distribution of the Indigenous Species of Phanerogamæ and Cryptogamæ Vasculares in St. Croix and the Virgin Islands—Continued.

	St. Croix only.	Virgin Islands only.	Common to both.	Total.
Styraceæ	....	1	... .	1
Ebenaceæ	....	1	....	1
Oleaceæ	....	....	2	2
Apocynaceæ	....	2	9	11
Asclepiadaceæ	1	3	3	7
Convolvulaceæ	3	7	24	34
Hydroleaceæ	....	....	1	1
Boraginaceæ	3	4	17	24
Solanaceæ	1	8	12	21
Scrophulariaceæ	2	1	3	6
Bignoniaceæ	1	2	6	9
Acanthaceæ	3	2	10	15
Gesneriaceæ	....	....	1	1
Labiatæ	1	2	9	12
Verbenaceæ	5	2	13	20
Myoporaceæ	....	....	1	1
Plumbaginaceæ	....	....	1	1
Phytolaccaceæ	1	....	4	5
Chenopodiaceæ	....	....	3	3
Amarantaceæ	....	2	13	15
Nyctaginaceæ	1	....	5	6
Polygonaceæ	4	1	3	8
Lauraceæ	2	3	4	9
Thymelæaceæ	....	1	....	1
Euphorbiaceæ	3	5	30	38
Urticaceæ	4	7	10	21
Aristolochiaceæ	1	....	1	2
Begoniaceæ	....	1	....	1
Piperaceæ	2	3	7	12
B.—MONOCOTYLEDONES.				
Alismaceæ	1	....	....	1
Hydrocharidaceæ	....	....	1	1
Potameæ	....	3	2	5
Aroideæ	1	5	3	9
Typhaceæ	....	....	1	1
Palmæ	....	....	2	2
Commelynaceæ	....	2	3	5
Graminaceæ	4	14	35	53
Cyperaceæ	5	15	13	33
Liliaceæ	1	....	7	8
Smilaceæ	1	1	....	2
Dioscoreaceæ	....	3	....	3
Bromeliaceæ	....	3	5	8
Scitamineæ	....	....	1	1
Orchidaceæ	1	12	2	15
C.—CRYPTOGAMÆ VASCULARES.				
Lycopodiaceæ	....	1	1	2
Filices	4	15	15	34
	98	215	568	881
Naturalized species	17	6	109	132
Total	115	221	677	1013

I. PHANEROGAMÆ.

A. DICOTYLEDONES.

DILLENIACEÆ.

1. Davilla rugosa, Poir.

St. Thomas (Griseb. Fl. p. 3).

ANONACEÆ.

2. Anona muricata, L. (v. Soursop, Susakka).

Fl. Feb.–May. Leaves with a peculiar strong scent, used against fever and vermin. Fruit edible; pulp resembling curdled milk, acidulous. In forests and thickets, common.—All islands.

3. A. laurifolia, Dun. (v. Wild Soursop).

Fl. Feb.–May. Resembling the former species in the foliage, but leaves of a quite different smell. Not uncommon in forests.—St. Croix; St. Thomas.

4. A. palustris, L. (v. Monkey-apple, Bunya).

Fl. May–June. Fruit not edible; used as bait for fishes. Common in marshy soil.—All islands.

5. A. squamosa, L. (v. Sugar-apple).

Fl. April–June. Foliage partly deciduous in March and April. Fruit edible, sweet, soft. Common in thickets.—All islands.

6. A. reticulata, L. (v. Custard-apple).

Fl. April–May. Fruit edible. In woods, not uncommon; also planted near dwellings.—All islands.—The enlarged top of the connective in all species of Anona is siliceous. None of the species enumerated above contains narcotic principles, as is the case with *A. Cherimolia*, Mill., and others.

7. Guatteria Ouregou, Dun.

St. Thomas (Griseb. Fl. p. 7).

8. Oxandra laurifolia, Rich. (*Uvaria excelsa*, Vahl in Hb. Juss.).

St. Croix (Caledonia Gut, West, p. 292).

23

MENISPERMACEÆ.

9. Cocculus domingensis, DC.

Fl. June–Aug. Stem woody, as much as two inches in diameter. Inflorescences often 3 or 4 uniserial in the same axil. (See Delessert, Icones, t. 96.) In forests, not common.—St. Thomas (near St. Peter, 1000').

10. Cissampelos Pareira, L. (v. Velvet-leaf). *a*) Pareira and *β*) microcarpa, DC.

Fl. Nov.–March. In forests and thickets, common.—All islands.

NYMPHÆACEÆ.

11. Nymphæa ampla, DC. (v. Water-lily). *β*) parviflora.

Fl. April–July. In rivulets.—St. Croix (Kingshill Gut); Vieques (Port Royal).

PAPAVERACEÆ.

12. Argemone mexicana, L. (v. Thistle).

Fl. the whole year. A very common weed in dry places.—All islands.

CRUCIFERÆ.

13. Nasturtium officinale, R. Br. (v. Water-cress).

Never seen flowering. Naturalized along rivulets.—St Croix; St. Thomas.

14. Sinapis brassicata, L. (v. Wild Mustard).

Fl. Jan.–June. Around dwellings and in waste places, not uncommon.—All islands.

15. Sinapis arvensis, L.

Fl. cleistogamous in February. Regular flowers later in the year. Naturalized; rare.—St. Croix (near Anguilla).

16. Lepidium virginicum, L.

Fl. the whole year. A common weed along roadsides and near dwellings.—All islands.

17. Cakile æqualis, L'Her.

Fl. Feb.–July. Rather common on sandy shores.—All islands.

[Cultivated species: *Brassica oleracea*, L. (v. Cabbage); *Lepidium sativum*, L. (v. Cress); and *Raphanus sativus*, L. (v. Radish).]

CAPPARIDACEÆ.

18. Cleome pentaphylla, L. (v. Massámbee).

Fl. the whole year. Flowers often polygamous. Leaves used as spinach. A common weed near dwellings and in waste places.—All islands.

19. C. puugens, W. (v. Wild Massánbee). *c*) and *β*) **Swartziana.**

Fl. the whole year. Common along roads and ditches.—All islands.

20. C. viscosa, L.

Fl. May–Dec. Naturalized here and there.—St. Croix; St. Thomas.

21. Moringa pterygosperma, G. (v. Horse-radish-tree).

Fl. the whole year. Root with a flavour of horse-radish. Naturalized and common near dwellings.—All islands.

22. Capparis amygdalina, Lam.

Fl. March–June. Leaves on young radical shoots linear in this and the two following species. Not uncommon in thickets.—All islands.

23. C. jamaicensis, Jacq. (v. Black Willie). *α*) **marginata** and *β*) **siliquosa.**

Fl. April–Aug. *α*) not uncommon; *β*) less common along the shore and in thickets.—All islands.

24. C. cynophallophora, L. (v. Linguan-tree). *α*) and *β*) **saligna.**

Fl. Feb.–Aug.—Glands 2–4, uniserial in the axils, exuding nectar when young before the time of flowering, and are to be considered as reduced branches or inflorescences.

25. C. verrucosa, Jacq.

Fl. April–May. A middle-sized tree. Not uncommon in forests on the Virgin Islands.

26. C. frondosa, Jacq. (v. Rat-bean).

Fl. Feb.–May. Seeds very poisonous. Common in forests.—All islands.

27. Morisonia americana, L. *α*) and *β*) **subpeltata,** Gris. in litt.

Fl. May–Oct. A considerable-sized tree. *α*) all islands; *β*) leaves subpeltate.—St. Croix (Spring Gut).

BIXACEÆ.

28. Bixa Orellana, L. (v. Roucon).

Fl. June–July. The red pigment of the fruit was generally used by the Caribs for anointing the whole body (Du Tertre). Naturalized in forests.—St. Croix (Crequis, Wills Bay); St. Thomas (Crown).

29. Trilix crucis, Griseb.

Fl. April–June. Stipules very variable. Petals always abortive in my specimens. A low tree or shrub. Uncommon in forests.—St. Croix (Wills Bay, Mt. Eagle); St. Thomas (Flag Hill); St. Jan (Cinnamon Bay).

30. Casearia sylvestris, Sw.

Fl. Jan.–Feb. and May–July. Seed covered by a red arillus. Common in forests and thickets.—All islands.

31. C. parvifolia, W. a) and β) **microcarpa,** Egg.

Fl. March–July. Flowers odorous. Stamens alternately of equal length. Not uncommon in forests. A low tree.—a) Virgin Islands; β) fruit small, 2′′′ diam., St. Croix.

32. C. ramiflora, Vahl. a).

Fl. Jan.–Feb. and July–Aug. Pedicel articulate below the middle. Arillus fibrous. Common in forests.—All islands.

33. Samyda glabrata, Sw.

Fl. June. Rare, in thickets on highest hill-tops.—St. Thomas (Crown, 1400′).

34. S. serrulata, L.

Fl. Feb.–May. Flowers odorous, precocious. Pedicels articulated at the middle. Leaves of young radical shoots linear. Common in thickets.—All islands.

VIOLACEÆ.

35. Ionidium strictum, Vent.

Fl. all the year round. Flower matutine. Rather uncommon in fissures of rocks in thickets.—St. Croix; Water Island.

TAMARICACEÆ.

36. Tamarix indica, Willd. (v. Cypress).

Fl. Sept.–Oct. Naturalized in gardens.—St. Croix; St. Thomas.

POLYGALACEÆ.

37. Polygala angustifolia, HB. Kth.

Fl. Dec.–Feb. In the shade of dense thickets. Rare.—St. Thomas (Cowell's Hill).

38. Securidaca Brownei, Gr. (*S. scandens* of West).

Fl. Feb.–April. Naturalized around Christiansted, v. Rohr.—St. Croix.

39. S. erecta, L.

St. Thomas (DC. Prodr. i, 341; Gris. Fl. p. 30).

40. Krameria Ixina, L.

Fl. July. The three narrow petals, resembling abortive stamens, are bent forward and cover the anthers. The two lateral ones are fleshy, and covered on the outer side with fleshy papillæ. Fruit 1-seeded by abortion. Gregarious along roadsides in dry localities, but uncommon.—St. Thomas (Bovoni).

CARYOPHYLLACEÆ.

I. PARONYCHIACEÆ.

41. Drymaria cordata, W. β) diandra.

Fl. May–June. In moist localities in the shade. Rare.—St. Croix (Spring Garden).

42. Cypselea humifusa, Turp.

Fl. July. Gregarious around a small fresh-water lagoon. Rare.— Water Island.

II. MOLLUGINEÆ.

43. Mollugo verticillata, L.

Fl. Aug. Leaves often fleshy. On rocky shores. Rare.—Buck Island, near St. Thomas.

44. M. nudicaulis, Lam.

Fl. Sept.–Dec. Not uncommon in moist localities.—St. Croix; Buck Island near St. Croix; St. Thomas.

III. PORTULACEÆ.

45. Talinum triangulare, W.

Fl. all the year round. Flower open till 11 A. M. Sepals of unequal size. The large one 1-ribbed, the smaller one 3-ribbed. Petals often yellow (as represented in Jacq. Stirp. Americ. t. 135). Rather uncommon. On rocks near the seashore.—St. Croix; St. Thomas.

46. T. patens, W.

Fl. all the year round. Flower open from 3 P. M. till sunset. Petals pale red or yellow (Bot. Mag. t. 1543). Root tuberous. Here and there in rocky situations.—St. Croix; St. Thomas.

47. Portulaca oleracea, L. (v. Purslane). α) macrantha, β) micrantha, Egg.

Fl. the whole year. Flower open till 10 A. M. α) brownish, 5 petals, as many as 25 stamens, corolla 6''' diam. β) green, 4 petals, 10–12 stamens, corolla 3''' diam. Both varieties common along roadsides and in open spots.—All islands.

48. P. quadrifida, L. (Mant. 78).

Fl. all the year round. Petals 4, yellow, 2''' long. Flower open from 11 A. M. till 3 P. M. Leaves opposite, clasping together towards evening. A common weed in gardens and along roads.—All islands.

49. P. pilosa, L.

Fl. all the year round. Often nearly glabrous. Roots tuberous. Petals red or yellow, large. Corolla up to 16''' diam., open only till 9 a. m. Seeds dark brown. Leaves adpressing themselves downward to the stem towards evening. Not uncommon. Along ditches and in grass-fields.—St. Croix; St. Thomas.

50. P. halimoides, L.

Fl. June–Dec. Common along roadsides and among rocks.—St. Croix; St. Thomas.

51. Sesuvium portulacastrum, L. (v. Bay-flower).

Fl. all the year round. Sepals rosy inside. Common on sandy shores.—All islands.

52. Trianthema monogynum, L.

Fl. all the year round. Branches always originating in the axil of the smaller leaf. Stamens 7–17. Sepals and stamens rosy or white. Common on rocky shores.—St. Croix; St. Thomas.

MALVACEÆ.

53. Malvastrum spicatum, Gris. (v. Hollow-stock).

Fl. all the year round. Flower expanding in the afternoon. Very variable. A common weed along roads and in fields.—All islands.

54. M. tricuspidatum, Asa Gray.

Fl. all the year round. Common along roads and ditches.—All islands.

55. Sida carpinifolia, L. c) and β) brevicuspidata.

Fl. Sept.-March. Pedicel geniculate at the base, or as often not so. Petals imbricate dextrorsely or sinistrorsely. Both forms very common weeds everywhere in dry localities.—All islands.

56. S. glomerata, Cav.

Fl. Aug.-Oct.—Buck Island near St. Thomas; Vieques.

57. S. ciliaris, L.

Fl. Sept.-March. Flower expanded till 10 A. M. Stipules always longer than the petioles. Leaves closely clasping the stem in the evening. Gregarious on roads and near ditches. Common.—All islands.

58. Sida jamaicensis, L.

Fl. Dec.–March. Flower expanded till 9 A. M. Calyx shorter than the corolla. In grass-fields and thickets. Often suffrutescent, 6' high. Common.—All islands.

59. S. spinosa, L. *a)*, *β)* **angustifolia, Lam.**, and *γ)* **polycarpa, Egg.**

Fl. Sept.–March. *γ)* suffrutescent, 4' high. Pedicel as long as the whole leaf. Pistils, ovaries, and carpids always 12. *a)* and *β)* common in grass-fields and pastures. *γ)* near rivulets.—All islands.

60. S. rhombifolia, L. (v. Swart Marån). *γ)* **retusa.**

Fl. Dec.–March. Petals showing a purple blot at the base. Common in waste places.—All islands.

61. S. tristis, Schlecht. (Linnæa, iii, 271).

St. Thomas (Schl.).

62. S. supina, L'Her. *a)* **glabra and** *β)* **pilosa, Egg.**

Fl. Nov.–March. Two very distinct forms : *a)* in shady, moist places; *β)* in dry localities. Not uncommon in thickets and forests.—All islands.

63. S. arguta, Cav. (not *S. arguta,* Sw., as stated in Griseb. Syst. Unters. p. 31)

St. Croix (West, 297); St. Thomas (Schl.).

64. S. nervosa, DC. *a)* **and** *β)* **viscosa, Egg.**

Fl. Dec.–April. *β)* viscous and glandular pilose. Petals reddish; pistils red. Not uncommon along roads and ditches.—All islands.

65. S. acuminata, DC. *a)* **macrophylla** and *β)* **microphylla.**

St. Thomas (Schl.). "In locis siccis."

66. S. cordifolia, L. *β)* **althæfolia, Sw.**

Fl. March. Here and there along roads.—St. Croix (West, 297); St. Jan (Bethania).

67. S. humilis, W. (?) **Cav.**

St. Thomas (Schl.). "In locis umbrosis."

68. Abutilon periplocifolium, G. Don. *a)* **and** *β)* **albicans, carpids 3-ovulate.**

Fl. all the year round. Seeds dimorphous. The two seeds in the superior cell glabrous, the one in the inferior silky. *a)* not uncommon along roads. *β)* uncommon.—St. Croix (*a* and *β*); St. Jan (*β*).

69. A. umbellatum, Sw.

Fl. Dec.–March. Seeds cordate, brown. Not very common in open, dry localities.—All islands.

70. A. indicum, G. Don (v. Mahoe). *a*) and *β*) **asiaticum.**

Fl. all the year round. Flower expanded after 3 P. M. only. Both forms common along roads and on waste places.—St. Croix; St. Thomas.

71. A. lignosum, Rich. (v. Marsh-mallow).

Fl. Nov.–May. Flower expanded during the afternoon only. Seeds irregularly triangular, verrucose, grey.—St. Croix.

72. Bastardia viscosa, Kth. *a*).

Fl. all the year round. Flower expanded during the afternoon only. Common along roads and in dry localities.—All islands.

73. Malachra capitata, L. *a*) and *β*) **alceifolia,** Jacq.

Fl. Dec.–March. Flower expanded only till 2 P. M. Along ditches and in moist places. *a*) rather common; *β*) less common.—All islands.

74. M. urens, Poit.

Fl. April. Petals yellow, puberulous externally. Seeds smooth, glabrous. Uncommon on waste places.—St. Thomas (western shore of the harbour).

75. Urena lobata, L. *a*) **americana.**

Fl. Nov.–June. Flower expanded till 10 A. M. In forests.—St. Croix (rare; Prosperity on the north coast); St. Thomas; St. Jan (not uncommon).

76. Pavonia spinifex, Cav.

Fl. Oct.–Dec. Rather common in thickets and forests.—All islands.

77. P. racemosa, Sw.

Fl. Oct. In marshy soil among Laguncularia and Conocarpus.—St. Croix (uncommon; Salt River).

78. Kosteletzkya pentasperma, Gr.

Fl. Aug. Flower expanded till 10 A. M. In marshy soil. Rare.—St. Thomas (Krumbay).

79. Abelmoschus esculentus. W. A. (v. Okro).

Fl. all the year round. Fruit used immature as a vegetable. Cultivated and naturalized near dwellings.—All islands.

80. Hibiscus clypeatus, L.

St. Croix (West, p. 298).

81. H. vitifolius, L.

Fl. Dec.–March. Along roads and in thickets.—St. Croix (naturalized in the eastern part of the island).

82. H. Sabdariffa, L. (v. Red Sorrel).

Fl. Oct.-Nov. Leaves used as a vegetable. Calyx at length fleshy, used for lemonade. Cultivated and naturalized here and there.—St. Croix; St. Thomas.

83. H. phœniceus, Jacq.

Fl. Sept.-March. Rather common in thickets, especially near dwellings.—St. Croix; St. Thomas.

34. H. brasiliensis, L.

St. Croix (West, p. 298)

85. Gossypium barbadense, L. (v. Cotton-tree). a) and β).

Fl. all the year round. Down stellate. Common in dry localities. Formerly cultivated.—All islands.

86. G. vitifolium, Lam.

Naturalized in St. Thomas (Schl.), perhaps from having been cultivated in former times.

87. Paritium tiliaceum, A. Juss. (v. Mahoe).

Fl. Oct.-March. Bark employed as rope. Along coasts, but rare.— St. Croix (West, p. 297); St. Thomas (Schl.); St. Jan (Fish Bay).

88. Thespesia populnea, Corr. (v. Otaheite Tree).

Fl. all the year round. Very easily propagated by cuttings. A shady tree with very hard wood. Naturalized and cultivated everywhere, especially in moist localities. All islands.

All Malvaceæ are protandrous.

[Cultivated species: *Althæ rosea*, L. (v. Hollyhock); *Hibiscus rosa-sinensis*, L. (v. Chinese rose); and *H. mutabilis*, L. (v. Changeable Hibiscus).]

BOMBACEÆ.

89. Adansonia digitata, L. (v. Guinea Tamarind).

Fl. June-July. Leaves deciduous in March-April. The acid pulp of the fruit used for lemonade. Naturalized in wooded valleys.—St. Croix (Prosperity; Crequis); St. Thomas.

90. Eriodendron anfractuosum, DC. (v. Silk-cotton-tree).

Fl. Feb.-April. Leaves deciduous March-April. Stem growing to immense size. Common in forests. All islands.

91. Myrodia turbinata, Sw.

St. Croix (Spring Garden, West, p. 298).

92. Helicteres jamaicensis, Jacq.

Fl. March–Aug.   Spiral of carpids 2½.   Common in thickets.—All islands.

BÜTTNERIACEÆ.

93. Guazuma ulmifolia, Lam. (v. Jackass Calalu).

Fl. April–June.   Wood used for oars.   Not uncommon in pastures.—St. Croix; St. Thomas.

94. Theobroma Cacao, L. (v. Cocoa-tree).

Fl. June.   Naturalized in shady valleys.—St. Croix (Prosperity; Mount Stewart).

95. Ayenia pusilla, L.

Fl. all the year round.   Flowers often transformed into a hollow monstrosity by the larva of a wasp.   Fruit muricate.   In thickets, common.—All islands.

96. Melochria pyramidata, L.

Fl. all the year round.   Common in pastures.—St. Croix.

97. M. tomentosa, L. (v. Broom-wood).

Fl. All the year round.   Calyx tomentose, greyish white.   Tomentum interspersed with glandulous hairs.   Used for brooms.   Common in dry thickets.—All islands.

98. M. nodiflora, Sw.

Fl. Nov.–July.   Common in pastures and along roads.—All islands.

99. Waltheria americana, L. (v. Marsh-mallow).

Fl. Oct.–May.   Common in pastures.—All islands.

TILIACEÆ.

100. Triumfetta Lappula, L. (v. Bur-bush).

Fl. Nov.–April.   Common in thickets.—All islands.

101. T. althæoides, Lam. (v. Mahoe).

Fl. Dec.–March.   In forests, uncommon.—St. Croix; St. Thomas.

102. T. semitriloba, L. (v. Bur-bush).

Fl. Oct.–March.   In thickets and along roads, common.—All islands.

103. T. rhomboidea, Jacq.

Fl. Dec.–April.   Uncommon in thickets.—St. Croix (Spring-gut).

104. Corchorus acutangulus, Lam.

Fl. June–Nov.   The lowest serratures of the leaves in my specimens often show one or two long setaceous bristles, as stated in DC. Prodr.

i, 505.   Griseb. Fl. p. 97, does not mention them, as he does in *C. olito-rius*, neither does the figure in Wight's Icones, iii, t. 739, show them in this species.   From observations made by me on *C. acutangulus*, as well as on *C. hirtus*, such bristles on the lower serratures of the leaves are of no specific value in this genus, being a variable feature.   In gardens and near dwellings, not uncommon.—St. Croix; St. Thomas.

105. C. siliquosus, L. (v. Papa-lolo).

Fl. Nov.–July.   Leaves used as a vegetable (Calalu).   Along roads and in pastures, common.—All islands.

106. C. hirtus, L.

Fl. June–Sept.   Two lowest serratures of the leaves sometimes showing one or two setaceous bristles.   In gardens and along roads, not uncommon.—St. Croix; St. Thomas.

107. C. hirsutus, L.

Fl. all the year round.   Hairs of the stem scabrous.   On sandy shores, common.—All islands.

TERNSTRÖMIACEÆ.

108. Ternströmia elliptica, Sw.

Fl. Feb.–April.   The two bracts at the base of the persistent calyx are to be considered as such (Swartz, Flora Ind. Occ. p. 961; DC. Prodr. i, p. 523; and Hook. & Benth. Genera Plant. i, p. 182), and not as sepals (Griseb. Fl. p. 103) on account of their being deciduous, but the sepals not.   The number of ovules in my specimens are about twenty in each cell.   (Hook.and Benth. l. c. ascribe to the genus only two, rarely three to six, in each cell; Grisebach l. c. only two to four.   In the Catal. Plant. Cub. p. 36, Griseb. mentions, however, a variety of *T. obovalis*, Rich., with ten to thirteen ovules in each cell.)   Sepals rosy, flowers fragrant   In forests on high hills, rare.—St. Croix (Maroon Hill, 900′); St. Jan (Bordeaux Hill, 1200′).

GUTTIFERÆ.

109. Clusia rosea, L. (v. Chigger-apple).

Fl. May–Sept.   Aërial roots as much as 20′ long, supporting the young trees on rocks or other trees.   In forests.—St. Croix (rare, Wills Bay); Virgin Islands (not uncommon).

110. C. alba, L. (v. Wild Mamey).

St. Croix (West, p. 312).   Probably a mistake for the first named species.

111. Mammea americana, L. (v. Mamey).

Fl. Feb. and later in Aug. Fruit generally one-seeded, eatable. Common in forests and planted along roads.—All islands.

112. Calophyllum Calaba, Jacq. (v. Santa Maria).

Fl. May–July. In forests along rivulets.—St. Croix (common in the northern part of the island); St. Thomas (rare).

CANELLACEÆ.

113. Canella alba, Murr. (v. White-bark).

Fl. Jan.–April. Berry dark crimson. Leaves used in warm baths for rheumatism. On sandy shores and in forests.—All islands.

ERYTHROXYLACEÆ.

114. Erythroxylum ovatum, Cav. (v. Wild Cherry, Brisselet).

Fl. April–Sept. Precocious. Branches, as a rule, transformed into brachyblasts. Common in thickets.—All islands.

(*E. areolatum*, West, p. 286, and *E. brevipes*, Bertero in Schlecht. Florula, are, no doubt, mistakes for the species mentioned above.)

MALPIGHIACEÆ.

115. Byrsonima spicata, Rich.

Fl. July–Aug. In forests, rare.—St. Croix (Parasol Hill); St. Thomas (Signal Hill); St. Jan (Bordeaux).

116. B. lucida, Rich.

Fl. Oct.—St. Thomas (DC. Prodr. i, 580); Vieques (Campo Asilo).

117. Bunchosia Swartziana, Gris.

Fl. July. Pedicel uniglandular and bibracteolate at the joint. Very much attacked by insects. In thickets.—St. Croix (rare, Kingshill); St. Thomas (not uncommon); St. Jan.

118. Galphimia glauca, Cav. (Icon. v, p. 61) (*G. gracilis*, Bartl.).

Fl. all the year round. Naturalized in gardens.—All islands.

119. Malpighia glabra, L. (v. Cherry).

Fl. May–June. Fruit edible. Common in thickets.—St. Croix; St. Thomas.

120. M. urens, L. a) and β) lanceolata.

Fl. June–Oct. a) common in thickets.—All islands; β) rare, St. Croix (Spring-gut).

121. M. Cnida, Spreng. (Neue Entdeck. iii, 51).

Fl. June–Sept. Along roads and in thickets, not uncommon.—St. Jan; Water Island; Vieques.

122. M. angustifolia, L.

Fl. June–Oct. In thickets, not uncommon.—Water Island; Vieques.

123. Stigmaphyllon periplocifolium, Juss.

Fl. all the year round. Samaræ red. In thickets, common.—All islands.

124. Heteropteris purpurea, Kth.

Fl. all the year round. Common in hedges and thickets.—All islands.

125. H. parvifolia, DC. (v. Bull Vis).

Fl. all the year round. As common as the preceding species.—All islands.

SAPINDACEÆ.

126. Cardiospermum Halicacabum, L. (v. Balloon-vine).

Fl. Sept.–March. Rather common in thickets and near dwellings.—St. Croix; St. Thomas.

127. C. microcarpum, Kth.

Fl. Jan.–March. In thickets, rare.—St. Croix (Spring-gut); St. Jan (Enigheit).

128. Serjania lucida, Schum. (v. White Vis, Cabrite rotting).

Fl. Dec.–June. Stem used as rope. Common in thickets.—All islands.—(*Paullinia curassavica,* West, p. 281, is no doubt a mistake for this species.)

129. Cupania fulva, Mart.

Fl. January. In forests, not uncommon.—Virgin Islands.

130. Sapindus inæqualis, DC. (v. Soap-seed).

Fl. Dec.–Jan. Seeds used for ornaments. In forests along rivulets. Not uncommon.—St. Croix.

131. Schmidelia occidentalis, Sw.

Fl. May–Sept. Not uncommon in forests, especially in St. Croix.—All islands.

132. Melicocca bijuga, L. (v. Keneppy tree).

Fl. April–May. Leafless during flowering. Flowers fragrant. Fruit astringent, edible. Naturalized and now very common everywhere, often forming a secondary growth in cleared woodland. Introduced from the Spanish main.—All islands.

133. Dodonæa viscosa, L.

Fl. April. On sandy seashores, rare.—St. Croix (Sandy Point).

MELIACEÆ.

134. Melia sempervirens, Sw. (v. Lilac, Hagbush).

Fl. all the year round. Common in forests and near dwellings.—All islands.

135. Trichilia hirta, L.

Fl. June–July. Common in thickets.—All islands.

(*Guarea trichilioides*, Jacq., said to occur in St. Croix (West, p. 281), seems to me rather doubtful.)

136. Swietenia Mahagoni, L. (v. Mahogany).

Fl. April–June. In wooded valleys and along roads and dwellings. Not uncommon.—St. Croix; St. Thomas.

GERANIACEÆ.

[Cultivated occur several species of Geranium, L'Her., and Pelargonium, L'Her.]

BALSAMINACEÆ.

137. Balsamina hortensis, Desp. (v. Lady-slippers).

Fl. all the year round. Naturalized everywhere in gardens. Seeds often germinating in the capsule.—All islands.

AURANTIACEÆ.

138. Citrus medica, L. α) (v. Citron). β) **Limonum,** Risso (v. Lime).

Fl. April–May. α) naturalized, but rare, in gardens. β) naturalized, common in gardens and near dwellings, also in forests.—All islands.

139. C. Aurantium, L. α) (v. Orange). β) **Bigaradia,** Duh. (v. Seville Orange).

Fl. May–July. Both forms naturalized in gardens, especially α). Common in St. Croix; rare in St. Thomas and St. Jan, where the species is said to have died out nearly, from disease.—(Mentioned also by Breutel, London Journal of Botany, ii.)

140. C. buxifolia, Padr. (v. Forbidden Fruit).

Fl. July. Naturalized in a few places.—St. Croix; St. Thomas.

141. C. decumana, L. (v. Shaddock).

Fl. July–Aug. Fruit used for preserves. Naturalized in gardens.— St. Croix; St. Thomas.

142. Triphasia trifoliata, DC. (v. Sweet Lime).

Fl. April–June. Naturalized in thickets and near dwellings. Common in all the islands.

[Cultivated species: *Murraya exotica*, L. (v. Cyprian), and *Cookia punctata*, Retz.]

OXALIDACEÆ.

143. Oxalis Martiana, Zucc.

Fl. May–Aug. Naturalized in gardens on all the islands.

144. O. corniculata, L. β) **microphylla,** Poir.

Fl. all the year round. Gregarious in fields.—St. Croix (Anually); St. Thomas.

ZYGOPHYLLACEÆ.

145. Tribulus cistoides, L.

Fl. all the year round. Along roads and in open spots, gregarious.— St. Croix (in the easternmost part of the island only).

146. T. maximus, L. (v. Centipee-root, Longlo).

Fl. all the year round. Stamens alternately of equal length. The whole plant is used in baths against boils. A very common weed along roads and in waste places.—All islands.

147. Guajacum officinale, L. (v. Lignum vitæ, Pockenholt).

Fl. March–April. Common in former times, but now nearly exterminated. On the seashore and in forests, rare.—All islands.

RUTACEÆ.

148. Pilocarpus facemosus, Vahl.

Fl. Feb.–March. Leaves undivided, 3-foliate or impari-pinnate in the same specimen (as stated in Hook. & Benth. Genera, i, 299, and Fl. Brasil. fasc. 65). Inflorescence terminal and axillary. A low tree. In forests, rare.—St. Jan (Kingshill, 1000′); Vieques (Ravn in Hb. Havn.). (Specimen from Montserrat in Hb. Havn. also named *P. laurifolius*, Vahl.)

149. Tobinia punctata, Gr.

Fl. Sept. Leaves often pinnate. Dots on the leaves pellucid. In thickets, not uncommon.—St. Croix.

150. T. spinosa, Desv.

Fl. May–June. Leaflets prickly on the principal nerves on both sides, bearing 2 stipular prickles at the base. Carpids 3 (2–1) globose, with a short beak, black, verrucose, 3′″ long. Seeds black, shining. Rare in forests.—St. Thomas (Flag Hill, 600′).

151. Fagara microphylla, Desf. (v. Ramgoat-bush) (*F. tragodes*, Jacq. in West).

Fl. June–Dec. Dots of the leaves pellucid. The whole plant has a strong smell. Not uncommon in thickets.—St. Croix; Buck Island, near St. Croix.

152. Zanthoxylum Clava-Herculis, L. (v. White Prickle).

Fl. April–June. Aculei corky, 6''' long, greyish, with a narrow brown point. In forests, not uncommon.—All islands.

153. Z. flavum, Vahl (Naturh. Selsk. Skrift. vi, 132, 1810) (v. Yellow Sander).

Not seen flowering. A fine timber-tree, used for furniture. Not uncommon in forests in former times, but now nearly extinct.—St. Jan (Bordeaux Hills) (St. Croix? St. Thomas?) (Montserrat, Ryan in Hb. Havn.); Martinique (West in Hb. Havn.).

154. Z. Ochroxylum, DC. (v. Yellow Prickle) (*Z. simplicifolium*, Vahl in Hb. Havn.).

Fl. June–Nov. ♀ Panicle 1'' long; pedicels ½''' long, bracteole at the base deciduous. Calyx 5-partite, ½''' diam. Petals 5, imbricate, white, ¾''' long, pellucid-dotted. Style thick, ¼''' high; stigmas triangular. Ovaries 3 on a short gynophore. Carpids 3 (1–2) globose, verrucose, partly dehiscent, 1½''' diam. Seed shining-black. Stem armed with large corky aculei, often connected and forming long ridges down the stem. ♪ Wood yellow. The whole plant is possessed of the same strong smell as Fagara. Not uncommon in forests.—St. Thomas (Flag Hill 600'); St. Jan (Rogiers) (Montserrat, Ryan in Hb. Havn.; Martinique, South America, Hb. Havn.). (A branch without flowers, marked *Z. macrophyllum*, St. Croix, Ryan in Hb. Havn., seems to belong to this species.)

155. Quassia amara, L. fil. (v. Quassia).

Fl. Nov.–Feb. Naturalized in gardens.—All islands.

156. Castela erecta, Turp.

Fl. Feb.–June. Petals purple. ♀ with 8 rudimentary stamens, alternately of equal size. Carpids 2–3–4. In dry thickets along the south coast, not uncommon.—St. Croix.

157. Picræna excelsa, Lindl. (v. Bitter-ash).

Not seen flowering. Wood very bitter, used for stomachic properties in drinks. In forests, rare.—St. Croix; St. Jan.

OLACACEÆ.

158. Schœpfia arborescens, R. S.

Fl. Feb.–March. Fruit nearly always 1-seeded by abortion. Here and there in forests.—St. Croix (Saltriver, Wills Bay); St. Thomas (Crown, 1400').

AMPELIDEÆ.

159. Cissus sicyoides, L. (v. Lambrali, Pinua koop).

Fl. all the year round. Flowers purple or yellow. Aërial roots long, filiform. Common in forests.—All islands.

160. C. trifoliata, L.

Fl. all the year round. On rocks and trees, not common.—St. Croix; St. Thomas.

161. C. acida, L.

Fl. June–Aug. In thickets near the coast, common.—All islands.

162. Vitis caribæa, DC.

Fl. June. In dense forests, rare.—St. Croix (Caledonia Gut); St. Thomas (Crown).

CELASTRACEÆ.

163. Maytenus elæodendroides, Gris. (Cat. Plant. Cub. p. 54). (*Rhamnus polygamus*, Vahl in Hb. Havn., and in West, p. 276.)

Fl. Dec. Flower brownish, small. Calyx 5-partite, ¾''' diam. Petals 5, oval, 1''' long. Stamens 5, often all or part of them transformed into petals and more or less sterile. Stigma subsessile, 2-lobed. Ovary 2-locular, 2-ovulate. Disc brown, undulate, ⅓''' high. Seed black with a red arillus. Rare in dry thickets.—St. Croix (Fair Plain).

164. M. lævigatus, Gris. in litt. (*Rhamnus lævigatus*, Vahl in Symb. Bot. iii, 41; *Ceanothus*, DC.).

Fl. May–Oct. Capsule tardily dehiscent, 1–3-seeded, 6''' long. Seeds brown, reticulate with red veins, 2''' diam. Arillus tough, white. A shrub or middle-sized tree. Not uncommon in forests.—All islands.

165. Elæodendron xylocarpum, DC. (v. Spoon-tree, Nut Muscat).

Fl. Sept.–Dec. Stamens often transformed, as in *Maytenus elæodendroides*. Drupe orange-coloured, 8''' long. Common on rocky shores; more uncommon in St. Croix.—All islands.

166. Myginda pallens, Sw.

Fl. Oct.–May. Common in thickets, principally in marshy soil.—All islands.

167. M. latifolia, Sw.

St. Croix (Pflug, sec. Vahl Symb. Bot. ii, 32); St. Thomas (Schl.).

168. Schæfferia frutescens, Jacq.

Fl. Sept.–Dec. Common in thickets.—All islands.

RHAMNACEÆ.

169. Reynosia latifolia, Gris. (Cat. Pl. Cub. 34) (v. Guama). Emend. in Eggers, Videnskab. Medd. fra Naturhist. Forening, Copenhagen, 1878, cum icone, p. 173. Fl. June–July. Common in dry thickets.—Virgin Islands.

170. R. mucronata, Gris. (l. c.) (Eggers, l. c). Not seen flowering. Rare in dry thickets near the coast.—St. Croix (easternmost part of the island, near Tague Bay).

171. Condalia ferrea, Gris. (v. Edden-wood). Fl. Sept.–Jan. Keel of the calyx-lobes foliaceous. Drupe oval, 2¼''' long. Not uncommon in thickets and forests.—All islands.

172. Colubrina ferruginosa, Brongn. Fl. Jan. and May–July. A low shrub. Common on sandy shores.—All islands.

173. C. reclinata, Brongn. (v. Snake-root, Mabee-bark). Fl. Nov.–March. Style 2–3-partite. Leaves used for the preparation of stomachic drinks. Not uncommon in thickets.—All islands.

174. Zizyphus reticulata, DC. (Prodr. ii, 20) (*Paliurus*, Vahl, Ecl. Am. iii, 6). Fl. July. Disc brownish. Capsule 3-locular, one seed in each cell, 5''' long, glabrous. Seeds purple; pulp reddish brown. In dry thickets, rare.—St. Croix (Fair Plain).

175. Gouania domingensis, L. (v. Soap-stick, Silvi). Fl. Oct.–Jan. Stem used as rope. Common in thickets.—All islands.

TEREBINTHACEÆ.

176. Bursera gummifera, L. (v. Turpentine-tree). Fl. April–Sept. Protandrous. Easily propagated by large cuttings, and generally used for forming fences. Common in forests and along roads.—All islands.

177. Hedwigia balsamifera, Sw. St. Croix (West in IIb. Havn. and p. 281 as *Icica altissima*).

178. Amyris sylvatica, Jacq. (v. Flamboyant). Fl. Feb.–April and July–Sept. Inflorescence trichotomous. Wood resinous and used for torches, especially in catching lobsters at night. Not uncommon in forests.—All islands.

179. Spondias lutea, L. (v. Hog-plum). Fl. March, coëtanous, and later July. Leaves deciduous in Feb. Fruit oval, edible. Common in forests.—All islands.

180. S. purpurea, L. (v. Jamaica Plum).

Fl. Feb.–March, precocious. Naturalized in gardens and wooded valleys.—All islands.

181. Rhus antillana, Egg. (n. sp.).

Sect. Sumach. Leaves impari-pinnate; leaflets 4–5-jugal, petiolulate, lanceolate, acuminate, obtuse at the base, entire, glabrous, chartaceous; veins prominulous beneath. Cyme ramose; branchlets bracteolate, equalling the leaves. Flower pedicellate, small, green, 5-merous, mostly ♂, the rest hermaphrodite. Calyx and petals persistent in the fertile flower. Stamens erect, a little longer than the petals, inserted into a fleshy central disc; filaments villous at the base. Ovary inserted upon a short fleshy gynophore. Drupe globose, glabrous, 1-seeded by abortion. A low tree. Approaching *R. metopium*, L. Fl. Jan. In forests, rare.—St. Thomas (Signal Hill, 1400′); St. Jan (Hb. Havn. as Xanthoxylum). (St. Croix, Stony-ground?)

182. Comocladia ilicifolia, Sw. (v. Prapra).

Fl. March–May. Root containing a lasting red dye. Common on limestone.—All islands.

183. Mangifera indica, L. (v. Mango-tree).

Fl. Feb.–April. Fruit edible. Introduced towards the close of last century, and now cultivated and naturalized everywhere.—All islands.

184. Anacardium occidentale, L. (v. Cashew, Cherry).

Fl. Dec.–April. Pedicel becoming fleshy, and containing in abundance a slightly astringent juice. Seeds used as almonds. Common in forests and along roads.—All islands.

LEGUMINOSÆ.

185. Crotalaria verrucosa, L.

Fl. all the year round. Naturalized along roads. Very common.—All islands.

186. C. retusa, L.

Fl. all the year round. Common along roads and in waste places. Naturalized.—All islands.

187. C. latifolia, L.

Fl. Nov. Leaves golden sericeous beneath. Corolla greenish. Not uncommon in thickets.—All islands.

188. C. incana, L. (v. Rattle-bush).

Fl. all the year round. Stipules deciduous, the scar exuding nectar afterwards, as well as the base of the bracteoles. Common along roads and near dwellings.—St. Croix; St. Thomas.

189. Indigofera tinctoria, L.

Fl. April–Aug. Cultivated in former times, but now only found wild or naturalized. Common in dry localities.—All islands.

190. I. Anil, L.

Fl. all the year round. The whole plant is much attacked by insects. Very common in dry thickets.—All islands.

191. Tephrosia cinerea, Pers. a) and β) litoralis, Pers.

Fl. Feb.–June. Both forms here and there in thickets.—All islands.

192. Cracca caribæa, Benth.

St. Croix (Schl.); St. Thomas (Gris. Fl. p. 183).

193. Coursetia arborea, Gris.

St. Jan (Gris. Fl. p. 183).

194. Sabinea florida, DC. (v. Waterpanna).

Fl. March–July. Precocious. Wood used for fishpots. Gregarious. Common in thickets and forests.—Virgin Islands. (Cultivated in St. Croix.)

195. Pictetia squamata, DC. (Prodr. ii, 314) (v. Fustic).

Fl. June. Flowering period only 5 or 6 days. Branches in this and the following species commonly transformed into brachyblasts. Common in forests and thickets.—Virgin Islands.

196. P. aristata, DC. (l. c.) (v. Fustic).

Fl. Feb., March, and June–Aug. Rather common in thickets.—Virgin Islands; St. Croix (Jacq. Hort. Schœnbr. ii, 60).?

(Both species are perhaps to be united, as proposed by Jacquin.)

197. Agati grandiflora, Desv.

Fl. all the year round. Naturalized in gardens, common.—All islands.

198. Sesbania sericea, DC.

Fl. Nov. In thickets near the coast, uncommon.—St. Thomas (Flag Hill).

199. Æschynomene americana, L.

Fl. Nov.–Jan. In pastures and along roads, not uncommon.—St. Croix.

200. Zornia diphylla, Pers.

Fl. July–Aug. In pastures on high hills, rare.—St. Thomas (Signal Hill, Crown).

201. Lourea vespertilionis, Desv.

Fl. Feb.–April. Naturalized in gardens.—St. Croix; St. Thomas.

202. Alysicarpus vaginalis, DC.

Fl. Nov.-Dec. Leaves very variable. Along roads, common.—All islands.

203. Desmodium triflorum, DC.

Fl. Dec.-Feb. Common near ditches and in moist localities.—All islands.

204. D. incanum, DC.

Fl. Oct.-Jan. Common in pastures.—All islands.

205. D. scorpiurus, Desv.

Fl. Dec.-Jan. In pastures, not very common.—St. Croix; St. Thomas (Duchass).

206. Desmodium tortuosum, DC.

Fl. Oct.-Jan. Common in pastures.—St. Croix; St. Thomas.

207. D. spirale, DC.

Fl. Nov.-Jan. Not uncommon in pastures and along roads.—All islands.

208. D. molle, DC.

Fl. Dec.-Jan. Lomentum often 3–4-jointed. Rather common in pastures.—St. Croix; St. Thomas.

209. Stylosanthes procumbens, Sw.

Fl. Oct.-Dec. Lomentum in my specimens always 2-jointed. Common along roads.—All islands.

210. S. viscosa, Sw.

St. Croix (West, p. 301). (Perhaps a mistake for the former species.)

211. Arachis hypogæa, L. (v. Pindars, Ground-nuts).

Fl. May-Aug. Seeds used for making cakes or eaten roasted. Cultivated and naturalized.—All islands.

212. Abrus præcatorius, L. (v. Jumbee-bead, Scrubber, Wild Liquorice).

Fl. Oct.-Feb. Leaves used for washing clothes. Common in thickets and on hedges.—All islands.

213. Rhynchosia minima, DC. *a*) and *β*) lutea, Egg.

Fl. all the year round. Seeds black, with small grey spots. *a*) Standard veined with purple; a low climber. *β*) Standard uniformly yellow; climbing up to 6′. Both forms common in pastures and thickets.—All islands.

214. R. phaseoloides, DC.

Fl. March. Stem laterally compressed. Rare in forests.—St. Thomas (Signal Hill, 1200′).

215. R. reticulata, DC.

Fl. all the year round. Leaflets as long as 1½″. Common on fences and along roads.—All islands.

216. Cajanus indicus, Spreng. (v. Pigeon-pea, Vendu bountje).

Fl. all the year round. Seeds used as a common vegetable for soup. Cultivated and naturalized.—All islands.

217. Clitoria Ternatea, L. (v. Blue Vine).

Fl. all the year round. Common in thickets.—All islands.

218. Centrosema virginianum, Benth. α) and β) **angustifolium.**

Fl. all the year round. Very common in ditches and on fences.—All islands.

219. Teramnus uncinatus, Sw., var. **albiflorus,** Egg.

Fl. Sept.–March. Corolla 1½′′′ long, constantly white. Legume 1″ long, black, pilose. Common in pastures and along roads.—St. Croix; St. Thomas.

220. Galactia filiformis, Benth.

Fl. Oct.–Jan. Roots often bearing small tubers. Common in thickets.—All islands.

221. G. tenuiflora, W. & A.

Fl. Feb.–June. In forests, rare. There seems not to be sufficient reason for uniting this species to the preceding, as done by Griseb. Fl. p. 194.—St. Thomas (Flag Hill); St. Jan (Rogiers).

222. Vigna luteola, Benth. (v. Wild Pea).

Fl. all the year round. Common in moist localities.—All islands.

223. Dolichos lablab, L. (*D. benghalensis,* Jacq.).

Fl. all the year round. Seeds brown. Very common along the seashores.—All islands.

224. Phaseolus lunatus, L. (v. Bonny Vis).

Fl. Dec.–Feb. Corolla white or rosy. Naturalized in thickets and near dwellings.—All islands.

225. Ph. vulgaris, L. (v. White Bean).

Fl. Feb.–July. Cultivated and naturalized near dwellings.—All islands.

226. Ph. alatus, L.

St. Croix (West, p. 299).

227. Ph. semierectus, L.

Fl. all the year round. Flower expanded only in the sun. Common along roads and in pastures.—All islands.

228. Canavalia parviflora, Benth. (Flor. Bras. xv, i, 177).

Fl. Feb. Inflorescence extra-axillary (as in *C. bonariensis*, Lindl. Bot. Reg. 1199). Legume broad on the back, without prominent ridges, 3″ long, 1¼″ broad. Seeds crimson, shining, ¾″ long. In forests, rare.—St. Thomas (Signal Hill, 1300′).

229. C. gladiata, DC. β) **ensiformis,** DC. (v. Sonr-eyes, Overlook) (*Dolichos acinaciformis,* Jacq. Icon. Rar. t. 559). Bot. Mag. 4027.

Fl. Aug.–Dec. Naturalized in provision grounds.—St. Thomas (Signal Hill, 1200′).

230. C. obtusifolia, DC. (*Dolichos rotundifolius,* Vahl).

Fl. all the year round. Common along the seashore.—All islands.

231. Mucuna pruriens, DC. (v. Cow-itch).

Fl. Oct.–Nov. In shady valleys. Rare.—All islands.

232. Erythrina Corallodendron, L. (v. Flamboyant).

Fl. Feb.–April. Precocious. Stamens all of unequal length. Rather common, especially along roads and near dwellings.—All islands.

233. E. horrida, Egg. (n. sp.).

Fl. Feb.–March. Very prickly. Approaching to the preceding, but stem, branches, petiole, and leaf-ribs on both sides armed with stout and straight prickles; legume terete, long-beaked. A low tree, branches procumbent. In forests, not uncommon.—All islands.

234. Piscidia Erythrina, L. (v. Dog-wood, Stink-tree).

Fl. March–April. Precocious. Only those individuals that flower drop the leaves. Common in thickets.—All islands

235. Drepanocarpus lunatus, Mey.

St. Croix (Isert, 1787, in Hb. Havn; West, p. 298).

236. Hecastophyllum Brownei, Pers.

Fl. June–Dec. Not uncommon on sandy shores.—All islands.

237. Andira inermis, Sw. (v. Dog Almond, Bastard Mahogany, Hon Kloot).

Fl. May–Aug. and Dec. Not uncommon in forests and along rivulets.—All islands.

238. Sophora tomentosa, L.

Fl. July–Jan. Along sandy shores, rare.—St. Croix (White's Bay, Turner's Hole).

239. Myrospermum frutescens, Jacq.

Fl. May–June. Legume resinous. Naturalized near dwellings.—St. Croix.

240. Hæmatoxylon campechianum, L. (v. Logwood).

Fl. Feb.–May. The young plants prickly on the stem. Here and there on sandy shores. More common in former times.—All islands.

241. Parkinsonia aculeata, L. (v. Horse-bean).

Fl. all the year round. Common in dry localities.—All islands.

242. Guilandina Bonduc, L. (v. Yellow Nickars).

Fl. May–Oct. Common along sandy shores.—All islands.

243. G. melanosperma, Egg. (n. sp.) (v. Black Nickars).

Fl. June–Oct. Resembling the preceding, but leaflets smaller, glabrous, shining, prickles red and seeds shining-black. Seeds used for ornaments. In dry thickets near the shore, rare.—St. Croix (Sandy Point, Grape-tree Bay).

244. G. Bonducella, L. (v. Grey Nickars).

Fl. all the year round. Anthers successively dehiscent. Flowers polygamous. Very common along sandy shores.—All islands.

245. Cæsalpinia pulcherrima, Sw. (v. Dudeldu).

Fl. June–Dec. Bracteoles large, subulate, but deciduous before the expansion of the flower. Commonly naturalized along roads and near dwellings.—St. Croix; St. Thomas.

246. Poinciana regia, Boj. (Bot. Mag. 2884) (v. Flamboyant).

Fl. May–July. Bracteoles as in the preceding. Leaves deciduous Dec.–April. A handsome tree of very quick growth. Naturalized in gardens and near dwellings.—St. Croix; St. Thomas.

247. Lebidibia coriaria, Schl. (v. Dividivi).

Fl. April–May. Legume used for tanning purposes. Rather common on dry hills.—Virgin Islands (St. Croix, cultivated).

248. Cassia Fistula, L.

Fl. Sept. Naturalized here and there in shady valleys.—St. Croix (The William).

249. C. grandis, L. (v. Liquorice-tree).

Fl. April–July. The pulp containing rhaphides in abundance. Naturalized and cultivated near dwellings.—St. Croix; St. Thomas.

250. C. bacillaris, L.

Fl. Nov.-May. Common in thickets and woods on high hills.—St. Thomas.

251. C. bicapsularis, L. (v. Stiverbush, Styver bla).

Fl. all the year round. Very common in waste places.—All islands.

252. C. florida, Vahl.

Fl. Dec. Naturalized near towns.—St. Thomas.

253. C. biflora, L. β) angustisiliqua, Lam.

Fl. Nov.-May. In thickets, rare.—St. Croix (Longford).

254. C. alata, L. (v. Golden Candlestick, Fleïti).

Fl. May-Nov. Along rivulets, not uncommon.—Virgin Islands (naturalized in St. Croix).

255. C. occidentalis, L. (v. Stinking-weed).

Fl. all the year round. Root used against fever. A very common weed near dwellings and in waste places.—All islands.

256. C. obtusifolia, L.

Fl. June-Nov. Common in dry localities.—St. Croix; St. Thomas.

(*C. triflora*, Vahl (Eclog. Am. iii, p. 11) (West, St. Croix), is a doubtful species. I have not been able to find the original specimen of Vahl in Hb. Havn.)

257. C. glandulosa, L. a) stricta, Schl., and β) ramosa.

Fl. all the year round. Both forms common in pastures and along roads.—All islands.

258. C. nicticans, L.

Fl. all the year round. In the same localities as the preceding.—St. Croix; St. Thomas.

259. Tamarindus indica, L. (v. Tamarind-tree).

Fl. March-June. Naturalized everywhere, especially near dwellings.—All islands.

260. Hymenæa Courbaril, L. (v. Locust-tree).

Fl. Jan. and July-Aug. Bracts large, early deciduous. The wood is an excellent timber on account of its being very hard and close-grained. In forests, here and there.—All islands.

261. Bauhinia tomentosa, L.

Fl. May-June. Leaves partly deciduous in March. Naturalized in gardens and near dwellings.—St. Croix; St. Thomas.

262. B. ungula, Jacq.
St. Thomas (Gris. Fl. 214).

263. Adenanthera pavonina, L. (v. Coquelicot).
Fl. July–Oct. Naturalized near dwellings and in shady valleys.—
All islands.

264. Neptunia pubescens, Benth.
Fl. Aug. Legume containing as many as 9 seeds. Rare.—Buck Island,
near St. Thomas.

265. Desmanthus virgatus, W. *a*) and *β*) **strictus**, Bert.
Fl. all the year round.   Both forms common in pastures and along
roads.—All islands.

266. D. depressus, Kth.
St. Thomas (Schl.).

267. Mimosa pudica, L. *a*) (v. Gritchee).
Fl. all the year round.   In pastures and along roads.—St. Croix (very
rare, Mt. Stewart); Virgin Islands (common).

268. M. asperata, L.
St. Thomas (Gris. Fl. 219).

269. M. Ceratonia, L. (v. Black Amaret, Amaretsteckel).
Fl. June–Dec.   On high hills.—St. Croix (West, p. 312; his specimens
are found in Hb. Havn.); Virgin Islands (common).

270. Leucæna glauca, Benth. (v. Wild Tamarind).
Fl. all the year round.   Leaflets closing together in strong sunlight.
Seeds used for fancy work, such as collars, baskets, etc. Very common
everywhere, also as secondary growth on cleared woodlands.—All islands.

271. Acacia Catechu, W.
Fl. May–July.   Stem furnished with strong black aculei.   Naturalized
in shady valleys.—St. Croix (Crequis).

272. A. nudiflora, W. (v. Amaret).
Fl. May and Nov.–Dec.   Protandrous.   Young foliage reddish.   Wood
used for fencing.   A low tree.   Common in thickets and woods.—Virgin
Islands.

273. A. sarmentosa, Desv. (v. Catch-and-keep, White Police).
Fl. July–Sept.   Stem generally angular or even winged.   A very spiny
climbing shrub, the recurved spines of which often make thickets impenetrable.   Common on dry hills.—Virgin Islands.

274. A. macracantha, IIB. β) **glabrens** (v. Stiuk Cashá).

Fl. Dec.-April. A shrub or low tree. Wood exhaling a very disagreeable odour. Common in thickets on dry hills.—All islands.

275. A. tortuosa, W. (v. Cashá).

Fl. all the year round. Flowers fragrant. Bracteoles rhomboid, ciliate.
Often gregarious. Common on dry hills.—All islands.

276. A. Parnesiana, W. (v. Cashá).

Fl. all the year round. Flowers fragrant; bracteoles spathulate, ciliate.
Foliage of this and the two former species eaten by goats, and their wood
generally used for making charcoal. Common in dry localities.—All
islands.

277. A. arabica, W.

Fl. Nov.-Jan. Naturalized near dwellings.—St. Croix; St. Thomas.

278. A. Lebbek, W. (v. Thibet-tree).

Fl. April-Sept. Leaves deciduous Nov.-March. Flowers fragrant.
Foliage eaten by cattle. The tree is often overgrown by *Loranthus emarginatus.* Naturalized in pastures and elsewhere.—St. Croix (very common); Virgin Islands (common, except St. Jan, where the tree seems not
to thrive).

(*A. frondosa,* W., var. *eglandulosa,* St. Thomas, is mentioned by
Schlechtendal as spontaneous, but, being an East Indian species, is most
probably only cultivated or at most naturalized. I have not seen the
species in the island.)

279. Calliandra portoricensis, Benth.

Fl. Feb. Climbing by the aid of young branches that twine themselves
around the branches of other trees. In forests, rare.—St. Jan (King's
Hill); Vieques.

280. C. purpurea, Benth. (v. Soldier-wood, West).

St. Croix (Gris. Fl. p. 224, probably on the authority of West. This
author, however, says, p. 312, that the tree is only cultivated in the isl-
and. His specimens are in existence in Hb. Havn. I have not seen the
tree on the island).

281. C. Saman, Gris. (v. Giant Thibet-tree).

Fl. May-Aug. A very large tree of quick growth. Naturalized near
dwellings and planted along roads.—St. Croix; St. Thomas.

282. Pithecolobium unguis-cati, Benth. *a*) and β) **forfex,** Kth. (v. Crab-prickle).

Fl. Sept.-Jan. Gynophore 1‴ long. Seeds black, shining; arillus
rosy. Wood used for fishpots. Both forms common on limestone and
in marshy soil.—All islands.

283. Inga laurina, W. (v. Lady-finger-tree).

Fl. July–Sept. and Jan.–March.   Petiole bearing a narrow wing on each side.   Corolla greenish.   (Jacquin's drawing does not show any wing on the petiole.   In the letterpress, however, of his Stirp. Am., he expresses a doubt whether the petiole is winged or not.)   Wood used for fences, etc.   Common in forests.—All islands.

[Cultivated species: *Pisum sativum*, L. (v. Green Pea); *Dolichos sphærospermus*, DC. (v. Black-eye Pea); *D. sesquipedalis*, L.; *Poinciana Gilliesii*, Hook.; and a *Casparea*.]

CHRYSOBALANACEÆ.

284. Chrysobalanus Icaco, L. (v. Cocoa-plum, Cacos).

Fl. Dec.–Feb. and July–Aug.   Fruit black or white; used for preserves. On sandy shores or in forests on high hills.   Common.—All islands.

ROSACEÆ.

[Many varieties of *Rosa gallica*, L., and *R. centifolia*, L., are cultivated in gardens on all the islands, and are flowering abundantly all the year round.   In the time of West (c. 1790), roses were rare, and flowered but seldom, so that we here seem to have an instance of gradual acclimatisation.]

MYRTACEÆ.

285. Calyptranthes Thomasiana, Berg (Linnæa, xxvii, 26).

St. Thomas (Ventenat and Ravn in Hb. Havn.).

286. C. Chytraculia, Sw.   β) **ovalis,** Berg, and ε) **Zyzygium,** Berg (l. c. p. 28).

In forests, rare.—St. Thomas; St. Croix.

287. C. pallens, Gris.

Fl. July–Aug.   Branchlets quadrangular.   In forests, rare.—St. Croix (Kingshill Gut); St. Thomas (Crown).

288. Myrcia coriacea, DC.   γ) **Imrayana,** Gris.

Fl. June–July.   In forests on high hills, uncommon.—All islands.

289. Jambosa malaccensis, DC.

Fl. April–May.   Naturalized in shady valleys; rare.—St. Croix (Crequis).

290. J. vulgaris, DC. (v. Pomerose-tree).

Fl. March–June.   Fruit used for preserves.   Naturalized along rivulets and in forests, common.—All islands.

291. Eugenia buxifolia, W.

Fl. June–Sept.   Petioles reddish.   Gregarious, especially along the seashore.—St. Croix; St. Thomas.

292. E. Poiretii, DC.

St. Thomas (Gris. Fl. 236).

293. E. monticola, DC.

Fl. July–Sept. Leaves variable, distichous. Flowers strongly fragrant. When not flowering, the shrub emits a fœtid smell. Rather common in forests.—All islands.

294. E. axillaris, Poir.

Fl. Aug.–Oct. Leaves variable. Petiole reddish. In thickets; rare.—St. Croix (Lebanon Hill, Fair Plain).

295. E. lateriflora, W. (*E. cordata*, DC. Prodr. iii, 272, and probably *E. sessiliflora*, ib. 273).

Fl. Sept.–Nov. Leaves very variable, ovate, cuneate, or oblong. Flowers sessile or subsessile, crowded in the axils. Berry globose, purple, 2''' diam. Common in thickets and forests.—All islands.

296. E. sessiliflora, Vahl (Symb. Bot. iii, 64).

Fl. July–Oct. Fruit large, rosy, $\frac{3}{4}''$–1'' diam. Flowers sessile, large, white, 5''' diam. In thickets, not uncommon.—St. Croix; St. Thomas (Cowell's Hill).

(Both DC. and Gris. seem to confound these two very distinct species, the flowers and fruits of which are highly different in most respects. DC. Prodr. iii, 273, says of his *E. sessiliflora:* Fructus dimidio minor quam *E. laterifloræ*, yet immediately above he says of this latter species: Fructus et sem. ignoti. Vahl's description is very correct, also, of the fruit, of which he says: Pruni magnitudine, globosus.)

297. E. flavovirens, Berg (l. c.).

St. Jan (Ravn in Hb. Havn.).

298. E. glabrata, DC. (Prodr. iii, 274).

St. Croix (Berg).

299. E. pallens, DC. (*E. nitida*, Vahl in Hb. Havn.) (v. Cromberry).

Fl. Sept.–Nov. Leaves shining. In forests, uncommon.—All islands.

300. E. acetosans, Poir. (DC. Prodr. l. c. 283).

St. Jan (in forests, Berg in Linnæa, xxx, 662); St. Croix (Mount Eagle, Richard).

301. E. virgultosa, DC.

Fl. April–July. Leaves variable. Common along the seashore and in forests.—All islands.

302. E. procera, Poir. (v. Black Cherry, Rock-myrtle) (*Myrtus cerasina*, Vahl in West, p. 290).

Fl. Feb. and Aug.–Nov. Flowers fragrant; fruit edible; a favourite food for wild pigeons. In forests, common.—All islands.

303. E. pseudopsidium, Jacq. (*E. Thomasiana*, Berg) (v. Bastard Guava, Christmas Cherry).

Fl. April–Dec. Flowers fragrant; fruit oval. A shrub or low tree. In forests, not uncommon.—All islands.

304. E. ligustrina, W.

Fl. April and Sept. In thickets and woods, common.—All islands.

305. E. portoricensis, DC. (Prod. iii, 266) (*Stenocalyx*, Berg).

St. Croix (ex Ilb. Vahlii in Hb. Berol.).

306. E. uniflora, L. (v. Surinam Cherry).

Fl. March–Aug. Fruit edible, acidulous. A middle-sized tree. Naturalized and planted in gardens.—St. Croix; St. Thomas.

307. E. floribunda, West (v. Guava-berry).

Fl. June–Aug. Berry black, globose, shining, 4′′′ diam., aromatic; used for preserves or put in rum. In forests, not uncommon.—All islands.

(*E. marginata* and *E. micrantha*, West, p. 290, are not mentioned in Vahl's Symb. Bot. pars iii, as stated, and are probably included in some of the species enumerated above.)

308. Anamomis punctata, Gris.

Fl. June. In forests, rare.—St. Croix (Maroon Hill, Wills Bay); St. Jan (Cinnamon Bay).

309. Pimenta vulgaris, W. & A. (v. Cinnamon-bush).

Fl. June–July. In forests, rare. An excellent timber tree.—St. Croix (Maroon Hill); Virgin Islands.

310. P. acris, W. & A. (v. Bay-leaf). *a*).

Fl. July–Aug. From the leaves the well-known bay-rum is distilled. In forests near the coast, not common.—St. Croix; Vieques.

311. Psidium Guava, Radd. (v. Guava). *a*).

Fl. all the year round. Fruit edible; also used for preserves. Very common, overrunning pastures and becoming troublesome in many places.—All islands.

312. P. cordatum, Sims. (v. Sperry Guava).

Fl. May–July. Fruit fragrant. In thickets on hills, not uncommon.—Virgin Islands.

313. Punica granatum, L. (v. Pomegranate).

Fl. April–Oct. Flowers crimson or yellow Fruit the same. Naturalized in valleys and near dwellings.—All islands.

314. Mouriria domingensis, Walp. (*Petaloma Mouriri*, Sw.).

St. Croix (Baudonius Gut, West, p. 285, and specimens in Hb. Havn.). [Cultivated species : *Myrtus communis*, L. (v. Myrtle), and *Couroupita guianensis*, Aubl. (v. Nutmeg).]

MELASTOMACEÆ.

315. Clidemia hirta, Don.

St. Thomas (Riedlé sec. Naudin, Ann. des sc. nat. 1853, xviii, p. 532).

316. C. spicata, DC.

Fl. June–July. In forests, not uncommon.—All islands.

317. C. rubra, Mart.

St. Thomas (Gris. Fl. p. 248; Finlay sec. Naudin, l. c.).

318. Diplochita serrulata, DC.

Fl. Feb.–May. Not uncommon in wooded valleys.—St. Croix; St. Thomas.

319. Tetrazygia elæagnoides, DC.

Fl. April–Aug. Common in forests and on high hills.—All islands.

320. Miconia argyrophylla, DC.

St. Thomas (Finlay sec. Naudin, Gris. Fl. p. 256).

321. M. impetiolaris, Don.

Leaves as long as 1½'.—St. Croix (West in Hb. Havn.); St. Thomas (Gris. Fl. p. 256; Bonpland sec. Naudin. Montserrat (Ryan in Hb. Havn.).

322. M. prasina, DC.

St. Thomas (Riedlé sec. Naudin).

323. M. lævigata, DC.

Fl. March–July. In forests, not uncommon.—All islands.

324. M. angustifolia, Gris.

Fl. March. A good-sized shrub, often gregarious on limestone.—St. Croix (Benzon in Hb. Havn.); Virgin Islands (not uncommon. Montserrat (Ryan in Hb. Havn.).

[Several of the species mentioned by Naudin as having been collected in St. Thomas I omit as being a rather doubtful habitat. These are : *Tshudya berbiceana*, Gris. (*Miconia*, Naud.); *Cremanium amygdalinum*, Gris. (*Ossœa*, DC.), and *Nepsera aquatica*, Naud.]

LYTHRARIEÆ.

325. Ammania latifolia, L.

Fl. Dec.–June.   Here and there in moist localities.—St. Croix (Lower Cove, Anna's Hope); St. Thomas (Flag Hill).

326. Antherylium Rohrii, Vahl (Symb. Bot. iii, 66) (v. Prickle-wood).

Fl. Oct.–March.   Precocious.   Petiole bibracteate above the middle. In marshy soil near the coast.—St. Croix (rare; Fair Plain, Stony Ground); Virgin Islands (common).

[Cultivated species: *Lawsonia inermis*, L. (v. Mignonette), and *Lagerströmia indica*, L. (v. Queen of Flowers).]

ONAGRACEÆ.

327. Jussieua suffruticosa, L.   a) **ligustrifolia, Kth.**

Fl. all the year round.   Here and there in moist places.—St. Croix (Crequis, Golden Rock); St. Thomas (Caret Bay).

RHIZOPHORACEÆ.

328. Rhizophora Mangle, L. (v. Mangrove, Mangelboom).

Fl. all the year round.   Gregarious along the shore of lagoons.—All islands.   (See Botaniska Notiser, 1877, Lund, and Vidensk. Medd. fra Naturhist. Forening in Copenhagen, 1877–78.)

COMBRETACEÆ.

329. Terminalia Catappa, L. (v. Almond-tree).

Fl. Jan.–April and Sept.   Naturalized in valleys and near dwellings.—St. Croix (common); Virgin Islands (rare).

330. Laguncularia racemosa, G. (v. White Mangrove).

Fl. all the year round.   Wood used for fishpots.   Common in salt-water lagoons.—All islands.

331. Bucida Buceras, L. (v. Gregory).

Fl. May–Aug.   A splendid timber tree.   Leaves often attacked by a fungus (*Erineum*, vide Kunze mycol. Hefte, ii, 148).   Flowers often transformed into long monstrosities (figured already in P. Browne's Jamaica, tab. 23).   Common in valleys and especially along the coast.—All islands.

332. Conocarpus erecta, L. (v. Button-wood).   a) and β) **procumbens, Jacq.**

Fl. all the year round.   Common along the coast and in lagoons.—All islands.

[Cultivated species: *Quisqualis indica*, L.]

CUCURBITACEÆ.

(Griseb. Flora, and Naudin: Annales des sc. nat. 1859, '62, '63, and '66.)

333. Momordica Charantia, L. *a*) and *β*) **pseudobalsamina** (v. Maid-apple).
Fl. Dec. and April–Aug. Common on fences and near ditches.—All islands.

334. Luffa cylindrica, Roem. (Syn. Mon. ii, 63) (*L. Petola,* Ser. Wight Icon. ii, t. 499) (v. Strainer-vine).
Fl. Oct.–Dec. Tendril 5-fid. Fruit brown, 4″ long. Naturalized on fences.—St. Croix; St. Thomas.

335. Cucurbita Pepo, L. *a*) (v. Pumpkin) and *β*) **Melopepo** (v. Squash).
Fl. May.–Nov. and Feb. Fruit used extensively as a vegetable. Naturalized and cultivated.—All islands.

336. Lagenaria vulgaris, Ser. *a*) (v. Gobie) and *β*) **viscosa,** Egg. (v. Bitter Gobie).
Fl. Sept.–Jan. The whole plant has a strong smell. Tendril 2-fid. *β*) leaves viscous, petiole biglandular near the top. Used as a blister. Not uncommon in waste places. *a*) on fences. Fruit used for goblets.— St. Croix; St. Thomas.

337. Melothria pervaga, Gris.
Fl. Dec.–April. In thickets, not uncommon.—All islands.

338. Cucumis Anguria, L. (v. Cucumber).
Fl. Jan.–March. Anthers glabrous in the bud, pilose after dehiscence, collecting the pollen. Berry used for soup and pickles. Common in pastures and on fences.—All islands.

339. Cephalandra indica, Naud. (l. c. 1866, p. 14) (*Coccinia,* W. & A.).
Fl. Dec.–June. Naturalized near dwellings and in shady valleys.— St. Croix.

340. Trianosperma graciliflorum, Gris. (*T. Belangerii,* Naud.).
Fl. Nov.–Jan. Leaf 3–5-lobed. Tendril often bifid. In forests, not uncommon.—All islands.

341. T. ficifolium, Mart. (Syst. nat. med. veg. Bras. 79) (*Bryonia,* Lam.).
Fl. March. In forests, not uncommon.—St. Thomas (Soldier Bay); St. Jan (West, p. 301).

342. Anguria trilobata, L.
St. Croix (Ham's Bluff, West, p. 305).

343. A. glomerata, Egg. (n. sp.).
Fl. Feb.–March and May–Aug. Root tuberous. Stem suffruticose, bark greyish. Leaves alternate, ovate-triangulate or 3-lobed, some-

times 3-partite, narrowly cordate at the base, denticulate, acuminate, scabrous above, whitish pubescent beneath. Tendril simple. ♀ flowers glomerate, sessile or subsessile, 8–20 in the glomerule. Calyx urceolate-cylindrical, small. Petals 5, orange-coloured or red, lanceolate, erect, 5''' long. Style bifid; stigmas thick, globose, obsoletely 2-lobed. Ovary 2-locular; ovules 3–8 in each cell. Berries densely glomerate, sessile or subsessile, oval, glabrous, striate, red, 8''' long. Seeds 3–8, urceolate-globose, verrucose, brownish, 2''' long. ♂ unknown. A high climber. Stem often ½'' diam. at the base, succulent. In forests, not uncommon.—St. Croix (Jacob's Peak, Claremont,); St. Thomas (Picaru Peninsula).

All Cucurbitaceæ are protogynous.

[Cultivated species: *Sechium edule*, Sw. (v. Choco); *Cucumis sativus*, L. (v. Mutton-cucumber); *C. Melo*, L. (v. Muskmelon), and *Citrullus vulgaris*, Schrader (v. Watermelon).]

PAPAYACEÆ.

344. Carica Papaya, L. (v. Papaw).
Fl. March–Aug. Stem often branched. Fruit used as a vegetable. Common near dwellings and in waste places. All islands.

PASSIFLORACEÆ.

345. Passiflora suberosa, L. (v. Pop, Indigo-berry).
Fl. Sept.–Dec. Common on rocks and fences.—All islands.

346. P. pallida, L.
Fl. Oct.–Dec. In forests, rare.—St. Croix (Wills Bay); St. Jan.

347. P. hirsuta, L. (*P. parviflora*, Sw.)
St. Croix (West, p. 30.

348. P. peltata, Cav.
St. Thomas (Schl.).

349. P. rubra, L.
Fl. Sept.–Feb. In forests and on rocks.—St. Croix (rare); Virgin Islands (common).

350. P. laurifolia, L. (v. Bell-apple).
Fl. all the year round. Leaf-margin glanduliferous. Berry fragrant, containing an edible pulp. In thickets on high hills (perhaps only naturalized) and cultivated.—All islands.

351. P. incarnata, L.
St. Croix (West, p. 304).

352. P. fœtida, L. (v. Love in the mist).

Fl. Sept.–Jan. Protandrous. On fences and near ditches, common.—St. Croix; St. Thomas.

[Cultivated species: *P. quadrangularis*, L. (v. Grenadilla), the berry of which is edible.]

TURNERACEÆ.

353. Turnera ulmifolia, L.

Fl. March–Oct. In waste places, common.—All islands.

354. T. parviflora, Benth.

Fl. Sept.–Dec. and Jan.–May. Leaves always eglandular; calyx not tomentose. Gregarious on rocky seashores, rare.—St. Thomas (Cowell's Hill); Buck Island, near St. Thomas.

CACTACEÆ.

355. Mamillaria nivosa, Link (Pfeiffer Enum. Cact. 1837, p. 11) (*M. tortolensis,* Hort. Berol.).

Fl. all the year round. Flower pale yellow; berry clavate, purple. Seeds brownish. On rocks near the seashore.—Buck Island and Flat Cays, near St. Thomas; Tortola (Pf.).

356. Melocactus communis, DC. (v. Pope's Head).

Fl. all the year round. Berry clavate, purple, $\frac{3}{4}''$ long. Seeds black, verrucose. Up to four feet high. On dry hills and rocks, especially near the shore.—All islands.

357. M. atrosanguineus, Hort. Berol.

St. Thomas (Pf. l. c. p. 44).

358. Cereus floccosus, Hort. Berol. (v. Dildo).

Fl. Oct.–July. Berry depressed globose, dark crimson, $1\frac{1}{2}''$ diam. Pulp red; seeds small, black. On dry hills in thickets, common.—All islands.

359. C. armatus, Otto.

St. Thomas (Pf. l. c. p. 81).

360. C. triangularis, Haw. (v. Chigger-apple).

Fl. July. Berry large, crimson, edible, 5'' long, oval. On trees and rocks in forests, not uncommon.—All islands.

361. C. grandiflorus, Haw. (v. Nightblooming Cereus).

Fl. May–July. Naturalized in gardens and near dwellings.—St. Croix; St. Thomas.

362. Opuntia curassavica, Mill. (v. Suckers).

Fl. all the year round. Berry purple, $\frac{3}{4}''$ long, clavate. Gregarious in dry localities, which are often rendered impenetrable by its presence. Very common.—All islands.

363. O. Tuna, Mill. (v. Prickly Pear).

Fl. all the year round. Berry ovate, crimson, edible. Seeds small, black. Used for fencing purposes. In dry localities, very common.— All islands.

364. O. horrida, Salm. (v. Bull-suckers).

Fl. all the year round. Flower reddish-yellow. In dry localities, common.—St. Croix; St. Thomas.

365. O. spinosissima, Mill.

Fl. all the year round. Spines white, 5–8 in each cluster, deciduous on the stem. Flower $\frac{3}{4}''$ diameter. Plant reaching 20′–25′ high. In dry thickets, common.—All islands.

366. O. tuberculata, Haw. (v. French Prickly Pear).

Fl. the whole year. Flower small, yellow. Branches used for poultices. Plant 10′–15′ high. Naturalized and planted near dwellings.— St. Croix; St. Thomas.

367. O. coccinellifera, Mill.

Fl. all the year round. Plant 15′–20′ high. On limestone, uncommon.—St. Croix; St. Thomas.

368. O. catocantha, Hort. Berol.

St. Thomas (Pf. l. c. p. 166).

369. Peireskia aculeata, Mill. (v. Surinam Gooseberry).

Fl. July. Fruit acidulous, edible. 'Naturalized and cultivated.—St. Croix; St. Thomas.

370. P. Bleo, HB. K.

Fl. all the year round. Sepals accrescent on the fruit. Naturalized and cultivated in gardens.—St. Croix; St. Thomas.

[Cultivated species: *Cereus peruvianus,* Tabem.; *C. monoclonos,* DC.; *C. repandus,* Haw., and *C. Phyllanthus,* DC.]

CRASSULACEÆ.

371. Bryophyllum calycinum, Salisb. (v. Wonderful Leaf).

Fl. Jan.-March.—Naturalized in dry localities, common, gregarious.— All islands.

ARALIACEÆ.

372. Panax speciosum, Willd. (Spec. Plant. iv, p. 1126).

Not seen flowering. Leaflets 8, of unequal size, the central ones largest. Margin slightly undulate and denticulate. Upper surface covered with distant and deciduous muricate hairs; tomentum on the lower surface deciduous. A low tree. In forests, very rare. St. Jan (King's Hill, 1000', on the northern slope of the hills). (Cuba, Porto Rico, Caracas.)

UMBELLIFERÆ.

373. Eryngium fœtidum, L.

Fl. Sept.–May. Biennial. Along rivulets and in moist places, rare. —St. Thomas (Caret Bay).

374. Anethum graveolens, L. (v. Dill).

Fl. March–Oct. Naturalized along roads and near dwellings.—All islands.

[Cultivated species: *Petroselinum sativum*, Hoffm. (v. Parsley); *Daucus Carota*, L. (v. Carrot); *Pimpinella Anisum*, L. (v. Anise); *Fœniculum vulgare*, Gærtn. (v. Fennel); *Anthriscus cerefolium*, L. (v. Chervil), and *Apium graveolens*, L. (v. Celery).]

LORANTHACEÆ.

375. Loranthus emarginatus, Sw. (v. Baas-fram-boom).

Fl. all the year round. Inflorescences uniserial. On trees, especially *Acacia Lebbek* and *Pisonia subcordata*. Common.—All islands.

376. Phoradendron flavens, Gris.

Fl. April–June. Seed compressed, green, with white bands. On *Pisonia subcordata*, rare.—St. Croix (Stony Ground).

CAPRIFOLIACEÆ.

[Cultivated occur: *Sambucus nigra*, L. (Fl. April–July), and *Lonicera Caprifolium*, L. (v. Honey-suckle).]

RUBIACEÆ.

377. Genipa americana, L.

Fl. July. In forests on high hills, rare.—St. Thomas (Crown); St. Jan (Rogiers).

378. Catesbæa parviflora, Sw.

Fl. Sept.–Dec. Fruit black, shining. In dry thickets, uncommon.— St. Croix (Fair Plain).

379. Randia aculeata, L. *a)* and *β)* **mitis.**

Fl. April–July. *a)* in dry thickets, *β)* in shady valleys. Common.—
All islands.

380. Hamelia patens, Jacq.

Fl. all the year round. 6'–15' high. In shady valleys, not uncommon.—All islands.

381. H. lutea, Rohr.

Fl. all the year round. In forests, uncommon.—St. Croix; St. Thomas.

382. Gonzalea spicata, DC.

Fl. May–Oct. In pastures on high hills, above 1000', not uncommon.—
Virgin Islands.

383. Exostemma caribæum, R. S. (v. Black Torch).

Fl. June–Dec. Common in thickets.—All islands.

384. Portlandia grandiflora, L.

Fl. June–Dec.—St. Thomas (DC. Prodr. iv, p. 405; Gris. Fl. p. 324); St.
Croix (cultivated).

385. Rondeletia pilosa, Sw.

Fl. all the year round. In thickets.—St. Croix (rare, near Cane Bay);
Virgin Islands (common).

386. Oldenlandia corymbosa, L.

Fl. Feb.–March. Seeds brown, minutely verrucose. In waste places,
rare.—St. Croix (Government House yard).

387. O. callitrichioides, Gris. (Pl. Wright. p. 506).

Fl. Dec. Flower expanded early in the morning and late in the afternoon. Gregarious among stones.—St. Croix (Government House).

388. Spigelia anthelmia, L. (v. Worm-weed).

Fl. all the year round. In open, moist localities, uncommon.—St. Croix;
St. Thomas.

389. Guettarda scabra, Lam.

Fl. June–Dec. Flower expanded towards evening. Drupe dark crimson, often 1-seeded by abortion. In woods, common.—All islands.

390. G. parvifolia, Sw.

Fl. July–Oct. In woods, not uncommon.—All islands.

391. Stenostomum lucidum, G.

Fl. Dec.–April. In forests, rare.—St. Croix; St. Thomas (Signal Hill).

392. Chione glabra, DC.

Not seen flowering. In forests, rare.—St. Croix (Fair Plain); St. Thomas (Soldier Bay).

393. Scolosanthus versicolor, Vahl.

Fl. Oct.–Dec. Pedicels often transformed into spines, as mentioned by DC. (Prodr. iv, 484). Leaves 2′′′–3′′′ long.—St. Croix (West and Ryan in IIb. Havn.); St. Thomas (rather common in thickets); Water Island.

394. Erithalis fruticosa, L. α) and β) odorifera, Jacq.

Fl. Oct.–March. Along the coast, not uncommon.—All islands.

395. Chiococca racemosa, Jacq.

Fl. March–Dec. In forests, common.—All islands.

396. Ixora ferrea, Benth.

Fl. Feb.–May and Nov.–Dec. Among rocks on high hills over 1200′, not uncommon.—St. Thomas (Crown).

397. Coffea arabica, L. (v. Coffee-tree).

Fl. May–July. Berry ripe Nov.–Dec. Naturalized in shady localities. Formerly cultivated on most estates on a small scale, principally in St. Jan.—All islands.

398. Faramea odoratissima, DC. (v. Wild Coffee).

Fl. June. In thickets on high hills.—St. Croix (West and Benzon in IIb. Havn.); Virgin Islands (not uncommon).

399. Psychotria glabrata, Sw.

Fl. June–Sept. Here and there in shady valleys.—All islands.

400. P. tenuifolia, Sw.

Fl. May. In thickets on high hills, rare.—St. Thomas (Crown, 1500′).

401. P. Brownei, Sprg.

Fl. June–Sept. In woods, common.—All islands.

402. P. horizontalis, Sw.

Fl. May–Dec. Along roads and in thickets, common.—All islands.

403. Palicourea Pavetta, DC. α) and β) var. rosea, Egg.

Fl. Feb. and Aug. β) corolla-lobes rosy, anthers bluish, and stem brownish. In forests, not uncommon.—β) all islands. α) St. Thomas (Signal Hill).

404. Morinda citrifolia, L. (v. Pain-killer).

Fl. June–Aug. Leaves used against headache. Naturalized in gardens.—St. Croix; St. Thomas.

405. Geophila reniformis, Cham. & Schl.

Fl. Dec.–Jan. and Aug. On the ground in dense woods, rare.—St. Thomas (Signal Hill, St. Peter); Vieques (Hb. Havn.).

406. Ernodea litoralis, Sw.

Fl. Dec.–May. Along sandy coasts, not uncommon.—All islands.

407. Diodia rigida, Cham. & Schl. (Linnæa, iii, 341).

St. Thomas (Schl.).

408. D. sarmentosa, Sw.

St. Thomas (Schl.).

409. Spermacoce tenuior, Lam. (v. Iron-grass). α) and β) **angustifolia,** Egg.

Fl. all the year round. β) leaves linear-lanceolate. In pastures and along roads. Both forms common.—All islands.

410. Borreria verticillata, Mey.

Fl. May–Oct. Suffruticose. In pastures on hills.—St. Croix (Hb. Havn.); St. Thomas (not uncommon on Crown).

411. B. stricta, Mey. (Primit. Fl. Essequib. p. 83).

Fl. Dec.–March. In pastures, here and there.—St. Croix (Parade Ground).

(*B. vaginata*, Ch. & Schl. (St. Thomas, Schl.), is a doubtful species (DC. Prod. iv, 551).)

412. B. parviflora, Mey.

Fl. March–June. Along roads and in forests.—St. Croix (Benzon in Hb. Havn.); St. Jan (Rustenberg, not uncommon).

[Cultivated species: *Ixora Bandhuca,* Roxb. (v. Burning Love), and *I. stricta,* Roxb.]

SYNANTHEREÆ.

413. Sparganophorus Vaillantii, G.

Fl. March–Sept. In moist localities, not uncommon.—St. Croix; St. Thomas (DC. Prod. v, 12).

414. Vernonia arborescens, Sw. α) **Swartziana,** β) **Lessingiana,** γ) **divaricata,** Sw.

Fl. May–Dec. In thickets, all three forms not uncommon.—All islands.

415. V. punctata, Sw.

Fl. all the year round. In thickets, common.—All islands.

416. V. Thomæ, Benth. (Vid. Medd. fra Nat. For. 1852, p. 66).

Fl. all the year round. In thickets, not uncommon.—St. Thomas.

417. Elephantopus mollis, Kth.

Fl. March–May. Head 4-flowered. In pastures, here and there.—All islands.

418. Distreptus spicatus, Cass.

Fl. Jan.–March. In pastures and along roads, common.—All islands.

419. Ageratum conyzoides, L.

Fl. Dec.–June. Achenium usually 4-gonous. Along roads and ditches, common.—All islands.

420. Hebeclinium macrophyllum, DC.

Fl. June–Sept. Achenium black, 3-gonous. In forests.—St. Croix (rare; Caledonia, Wills Bay); St. Thomas (not uncommon).

421. Eupatorium odoratum, L. (v. Christmas-bush).

Fl. Nov.–March. Along roads and in thickets, common.—All islands.

422. E. repandum, W.

Fl. Dec.–July. On hills, not common.—All islands.

423. E. atriplicifolium, Vahl (Symb. Bot. iii, 96).

Fl. Dec.–May. Leaves coriaceous, glabrous; glandular impressions numerous on the upper surface. Flower odorous. On sandy shores, common.—All islands.

424. E. canescens, Vahl.

Fl. Oct.–Nov. In thickets, uncommon. St. Croix (Spring-gut); St. Thomas (DC. Prod. v, 155).

425. E. Ayapana, Vent.

St. Croix (naturalized sec. Vahl, who received it from Pflug; probably only cultivated).

426. E. cuneifolium, Willd.

St. Thomas (DC. Prod. v, 177).

427. Mikania gonoclada, DC.

Fl. Dec.–March. In forests.—St. Croix (rare; Caledonia); Virgin Islands (not uncommon).

428. Erigeron cuneifolius, DC. (Prod. v, 288).

Fl. Dec.–July.—Rhizome perennial, for which reason this species must be considered sufficiently distinct from the annual *E. Jamaicensis,* Sw. The two species are united into one by Prof. Grisebach in his Fl. p. 365. In pastures on high hills, not uncommon above 1200′.—Virgin Islands.

429. E. spathulatus, Vahl.

Fl. April–July. Along roads and ditches, rather common.—All islands.

430. E. canadensis, L.

Fl. June–Nov. Ray-flowers often ligulate. Along roads, common.—All islands.

431. Baccharis Vahlii, DC. (Prod. v, 411) (*B. dioica*, Vahl).

Fl. all the year round. As much as 30' high. On rocky seashores, gregarious, not uncommon. (The specific name of DC. is to be preferred to that of Vahl, notwithstanding the priority of the latter, for the reasons stated in the Prodromus.)—St. Croix (northwestern coast).

432. Pluchea odorata, Cass. (v. Sweet Scent, Ovra bla).

Fl. Feb.–April. Leaves used as tea against colds and as diuretic medicine. In moist localities, not uncommon.—All islands.

433. P. purpurascens, DC.

Fl. all the year round. Along rivulets, not uncommon.—St. Croix (Gallows Bay, Kingshill Gut).

434. Pterocaulon virgatum, DC.

Fl. all the year round. On dry hills, common.—All islands.

435. Melampodium divaricatum, DC. (Prod. v, 520) (*M. paludosum*, Kth.).

Fl. Oct.–Feb. Along ditches, gregarious, rare.—St. Croix (Jolly Hill).

436. Ogiera ruderalis, Gris.

Virgin Islands (Gris. Fl. p. 369).

437. Acanthospermum humile, DC.

Fl. all the year round. Leaves not glandular beneath. A common weed along roads.—St. Thomas.

438. Xanthium macrocarpum, DC. (Prodr. v, 523) (*X. orientale*, L.).

Fl. Oct.–Feb. A common weed, naturalized around dwellings.—All islands.

439. Parthenium Hysterophorus, L. (v. Mule-weed, White-head-broom).

Fl. all the year round. A very common weed everywhere.—All islands.

440. Ambrosia artemisiæfolia, L. β) **trinitensis.**

Fl. Sept.–Oct. Naturalized in waste places.—St. Croix (Fredrikssted).

441. Zinnia multiflora, L. (v. Snake-flower).

Fl. Feb.–Aug. Along roads, not uncommon.—Virgin Islands.

442. Z. elegans, Jacq.

Fl. May–Oct. Naturalized in gardens.—All islands.

443. Eclipta alba, Hassk.

Fl. June–Feb. In moist localities, not uncommon.—All islands.

444. Borrichia arborescens, DC.

Fl. all the year round. On sandy shores, gregarious.—St. Croix (common); St. Thomas (Smith's Bay).

445. Wedelia carnosa, Rich.

Fl. June–Jan. Along ditches, gregarious.—St. Croix (western part of the island, not uncommon).

446. W. buphthalmoides, Gris. (v. Wild Tobacco). *a*), *β*) **antiguensis, Nichols,** and *γ*) **dominicensis.**

Fl. all the year round. Leaves delicately fragrant. *a*) rare; *β*) and *γ*) common along roads and in thickets.—All islands.

447. W. affinis, DC. (Prod. v, 541) (*W. calycina*, Rich.).

St. Thomas (Wydler).

448. W. acapulensis, HB. K.

St. Thomas (Schl. in Linnæa, 1831, 727).

(Grisebach, Fl. 372, thinks these two species to be included probably in *W. frutescens*, Jacq.)

449. W. cruciana, Rich.

St. Croix (DC. Prodr. v, 542).

450. W. discoidea, Less. (Linnæa, 1831, 728).

St. Thomas (Less. l. c.).

451. Melanthera deltoidea, Rich.

St. Thomas (Less.).

452. Sclerocarpus africanus, Jacq. (Icon. Rar. i, t. 176).

Fl. Nov.-Dec. Along roads and in thickets, rare. (Naturalized?)—St. Thomas (Parade ground).

453. Bidens leucanthus, W.

Fl. Sept.-Dec. Under trees, on high hills.—St. Croix (West, p. 303); Virgin Islands (common).

454. B. bipinnatus, L.

Fl. Sept.-March. Achenium often 5-aristate. In pastures and along ditches, common.—All islands.

455. Cosmos caudatus, Kth.

Fl. Dec.-March. Along roads and in fields, not uncommon.—All islands.

456. Verbesina alata, L.

Fl. Feb.-Aug. Naturalized in gardens.—St. Croix; St. Thomas.

457. Synedrella nodiflora, G. (v. Fatten barrow).

Fl. all the year round. A common weed everywhere.—All islands.

458. Pectis punctata, Jacq.

Fl. Oct.–March. In pastures and along ditches, common.—All islands.

459. P. linifolia, Less.

St. Thomas (Less. Gris. Fl. p. 378).

460. P. humifusa, Sw.

Fl. all the year round. Gregarious on rocks and between stones, not uncommon.—All islands.

461. Egletes domingensis, Cass. α) glabrata, DC.; β) carduifolia, DC.; γ) genuina.

Fl. all the year round. On the sandy seashore, α) and γ) rather common. β) found by Oersted (Vid. Medd. 1852, p. 106).—St. Thomas.

462. Erechthites hieracifolia, Raf. α) and γ) cacaloides, Less.

Fl. all the year round. In moist localities, not uncommon.—St. Croix (γ); St. Thomas (α).

463. Emilia sonchifolia, DC.

Fl. Jan.–Oct. In shady localities. Naturalized, common.—All islands.

464. E. sagittata, DC. (Prodr. vi, 302) (*Cacalia coccinea*, Sims.).

Fl. all the year round. Naturalized in gardens.—St. Croix; St. Thomas.

(*Cacalia coccinea*, Sims., is, according to DC. Prodr. vi, 332, a synonym for *Emilia coccinea*. This latter species does, however, not occur in the Prodromus at all, and on a former page, 302, the *Cacalia* of Sims. is given as synonymous with *E. sagittata*.)

465. Leria nutans, DC.

Fl. June–March. In shady localities on hills, not uncommon.—All islands.

466. Brachyrhamphus intybaceus, DC. (Jacq. Icon. Rar. i, t. 162).

Fl. all the year round. Near dwellings and in waste places, a common weed.—All islands.

467. Sonchus oleraceus, L. (v. Wild Salad).

Fl. all the year round. Achenium mostly 4-furrowed. Along roads and near dwellings, common.—All islands.

(*Chrysogonum dichotomum*, sp. nov., Vahl, mentioned in West, p. 303, as occurring in St. Croix, is not described in any of Vahl's publications;

and as no specimens are to be found in Hb. Havn., I have not been able
to identify the species.)

[Cultivated species: *Helianthus annuus*, L. (v. Sunflower); *Pyrethrum
indicum*, Cass.; *Aster chinensis*, L.; *Tagetes patula*, L.; *Tithonia speciosa*,
Hook.; *Georgina variabilis*, Willd., and *Lactuca sativa*, L. (v. Salad).]

LOBELIACEÆ.

468. Isotoma longiflora, Prsl.

Fl. all the year round.   The whole plant is poisonous.   In shady localities and in pastures on high hills.   St. Croix (rare, Mount Pleasant,
Wills Bay); Virgin Islands (rather common on the hills).

GOODENOVIACEÆ.

469. Scævola Plumieri, L.

Fl. Jan.–April.   On sandy shores.—St. Croix (not uncommon); St.
Thomas (Smith's Bay).

MYRSINACEÆ.

470. Ardisia coriacea, Sw.

Fl. June–Aug.   Leaves minutely spotted beneath.   In forests and on
high hills, not uncommon.—All islands.

471. Jacquinia armillaris, L.   α) and β) **arborea**, V. (v. Bay Sallie).

Fl. Sept.–Feb.   On the rocky shore, not uncommon.—All islands.

SAPOTACEÆ.

472. Chrysophyllum Cainito, L. (v. Star-apple).

Fl. May–July.   Fruit edible.   In forests, rare.—St. Croix (Springfield);
St. Thomas (Signal Hill).

473. C. pauciflorum, Lam.

Fl. June.   In forests, uncommon.—St. Thomas (Flag Hill).

474. C. oliviforme, Sw.   β) **monopyrenum**.

Fl. July.   In forests, not very common.—St. Croix; St. Thomas.

475. C. microphyllum, Jacq. (v. Palmér).

Fl. Sept.–Jan.   In wooded valleys, rare.—St Croix (Bugby Hole); St.
Thomas (Santa Maria Gut).

476. C. glabrum, Jacq.

Fl. Sept.–Dec. and March–July.   In woods and thickets, common.—
All islands.

477. Sapota Achras, Mill. (v. Mespel).

Fl. Sept.–Oct. and March.   Fruit sweet, edible.   In forests and cultivated, common.—All islands.

478. S. Sideroxylon, Gris. (v. Bully wood).

Not seen in flower. A tall tree, affording a splendid purple, very hard timber. In forests, rare.—St. Jan (Baas Gut).

479. Sideroxylon Mastichodendron, Jacq. (v. Mastic).

Fl. Aug.-Sept. An excellent timber tree. In forests, rare.—St. Croix (Lebanon Hill); St. Thomas (Northside Bay); St. Jan (Baas Gut) (Montserrat, Ryan in IIb. Havn.).

480. Dipholis salicifolia, DC.

Fl. Feb.-March. In thickets and forests.—St. Croix (not uncommon in the western part of the island); St. Jan (Klein Caneel Bay).

481. Bumelia cuneata, Sw. (v. Break-bill).

Fl. Feb.-April. Branches often transformed into long spines. Very good timber tree. Along the coast principally in marshy soil, not uncommon.—All islands.

482. Lucuma multiflora, DC. (*Achras macrophylla*, Vahl in IIb. Havn.).

Fl. June-July and Dec.-Jan. Leaves as much as 1½′ long.—St. Croix (Hb. Havn. from Wills Bay); St. Thomas (here and there in forests; Signal Hill, 1500′).

STYRACEÆ.

483. Symplocos martinicensis, Jacq.

Fl. March-Aug. In forests on high hills. Flowers fragrant.—St. Thomas (Signal Hill above 1200′, not uncommon).

EBENACEÆ.

484. Maccreightia caribæa, A. DC.

Vieques (Duchassaing sec. Gris. System. Unters. p. 91).

OLEACEÆ.

485. Linociera compacta, R. Br.

Fl. May-Oct. In forests, rather common.—St. Croix; St. Thomas.

486. Forestiera porulosa, Poir. a) and β) **Jacquinii,** Egg. (Jacq. Ic. Rar. t. 625).

Fl. Feb. and Sept.-Oct. In thickets near the coast, uncommon.—a) St. Thomas (Cowell's Hill); β) St. Croix (northern shore near Claremont).

JASMINACEÆ.

487. Jasminum pubescens, W. (v. Star Jessamine).

Fl. all the year round. Naturalized in gardens.—All islands.

[Cultivated species: *J. officinale*, L.; *J. revolutum*, L. (v. Nepaul Jessamine), and *Nyctanthes Sambac*, L. (v. Double Jessamine).]

APOCYNACEÆ.

483. Thevetia neriifolia, Juss. (v. Milk-bush).

Fl. all the year round. Wood employed for building boats. In thickets on dry hills, common.—All islands.

489. Rauwolfia nitida, L. (v. Milk-tree).

Fl. all the year round. In forests and thickets, common.—All islands.

490. R. Lamarckii, A. DC. (v. Bitter-bush).

Fl. all the year round. On dry hills, common.—All islands.

491. Nerium Oleander, L. (v. Nerium).

Fl. all the year round. Naturalized in gardens and near dwellings. Common.—All islands.

492. Tabernæmontana (citrifolia, Jacq.?).

Fl. June–Aug. In thickets, here and there.—St. Thomas (Frenchman's Bay).

493. Vinca rosea, L. (v. Church-flower).

Fl. all the year round. Near houses and on waste places, very common.—All islands.

494. Plumieria rubra, L. (v. Red Franchipani).

Fl. all the year round. Naturalized near dwellings.—All islands.

495. P. obtusifolia, L. (v. White Franchipani).

Fl. all the year round. Naturalized in gardens.—All islands.

496. P. alba, L. (v. Snake-root, Klang hout).

Fl. all the year round. On rocks near the shore and in dry thickets, common.—All islands.

497. Echites agglutinata, Jacq.

Fl. July–Aug. In thickets, rare.—St. Croix (Cane Bay); St. Thomas (Flag Hill).

498. E. circinalis, Sw.

Fl. Dec. In forests, rare.—St. Thomas (Flag Hill).

499. E. neriandra, Gris.

Fl. Oct.–Jan. Here and there in thickets, not uncommon.—All islands.

500. E. suberecta, Jacq.

Fl. May–Aug. In thickets, uncommon.—St. Thomas (Cowell's Hill); St. Croix (West, p. 277).

501. E. barbata, Desv.

St. Croix; St. Thomas (DC. Prodr. viii, 453).

[Cultivated species: *Allamanda cathartica*, L., and *Tabernœmontana capensis*, L. (v. Cape Jessamine).]

ASCLEPIADACEÆ.

502. Metastelma parviflorum, R. Br.

St. Thomas (Duchass).

503. M. Schlechtendalii, Decs. (*M. albiflorum*, Gris.).

Fl. all the year round. In dry thickets, very common.—All islands.

(The specific distinction of Grisebach's species does not seem to be sufficiently permanent to justify a separation into two.)

504. Asclepias curassavica, L. (v. Wild Ipecacuana).

Fl. all the year round. Root used as an emetic. Along roads and ditches, common.—All islands.

505. A. nivea, L.

St. Thomas (Gris. Fl. 419).

506. Sarcostemma Brownei, Mey.

St. Thomas (West, p. 278, as *Asclepias viminalis*, Sw.).

507. Calotropis procera, R. Br. (v. Silk Cattún).

Fl. all the year round. Naturalized in dry localities, common.—All islands.

508. Ibatia muricata, Gris.

Fl. all the year round. In dry thickets, common.—All islands.

509. Fischeria scandens, DC.

Fl. Aug. In forests, rare.—St. Croix (Spring-gut).

[Cultivated species: *Hoya carnosa*, R. Br. (v. Wax-flower) and *Stephanotis floribunda*, A. Brongn.]

CONVOLVULACEÆ.

510. Ipomæa bona-nox, L.

Fl. Oct.-May. Naturalized in gardens.—St. Croix; St. Thomas.

511. I. Tuba, Don.

Fl. all the year round. On shrubs near the coast, uncommon.—All islands.

512. I. tuberosa, L.

Fl. Feb.-March. In forests, rare.—St. Croix (Bugby Hole); St. Thomas (Schl.).

513. I. dissecta, Pursh (v. Noyau Vine).

Fl. Nov.-May. Corolla-tube purple inside. The whole plant has a taste of prussic acid, and is used for the preparation of a liquor called Noyau. On fences and along roads, common.—All islands.

514. I. pentaphylla, Jacq.

Fl. Dec.-March. In thickets and along ditches.—St. Croix; St. Thomas.

515. I. quinquefolia, Gris.

Fl. Dec.-Jan. Corolla expanded from 8 A. M. to 3 P. M. In pastures and low thickets, common.—St. Thomas.

516. I. Batatas, Lam. (v. Sweet Potato). a), β) leucorrhiza, and γ) porphyrorhiza.

Fl. all the year round. Propagated by cuttings.. A common vegetable. Cultivated and naturalized everywhere.—All islands.

517. I. fastigiata, Swt. a).

Fl. Oct.-Jan. In thickets, not uncommon.—St. Thomas.

518. I. violacea, L. (v. Granni Vine).

Fl. Dec.-Feb. Coralla expanded towards evening. In forests and along rivulets, not uncommon.—All islands.

519. I. carnea, Jacq.

St. Croix (Wills Bay see. West, p. 272).

520. I. leucantha, Jacq. (Icon.Rar. ii, t. 318).

Fl. March-May. Capsule pilose; roots tuberous. On dry hills, not uncommon.—St. Jan (near Klein Kanelbay).

521. I. triloba, L. a) and β) **Eustachiana, Jacq.**

Fl. Sept.-March. Corolla expanded till 10 A. M. Both forms in moist localities, not uncommon.—St. Croix; St. Thomas.

522. I. umbellata, Mey.

Fl. Jan.-March. Along rivulets and ditches,.common.—All islands.

523. I. pes-capræ, Sw. (v. Bay Vine).

Fl. all the year round. Corolla sometimes white. On sandy seashores, very common.—All islands.

524. I. asarifolia, R. S.

Danish islands (Gris. Fl. p. 471).

(As this species is a native of Senegal, I doubt the correctness of the above habitat.)

525. I. quinquepartita, R. S. (*Conv. ovalifolius*, West (non Vahl) sec. DC. Prodr. ix, 367).

St. Croix (West, p. 271).

526. I. triquetra, R. S. (*Conv. triqueter*, Vahl, Symb. Bot. iii, 32).

St. Croix (West, p. 271); St. Thomas (Schl.).

527. I. repanda, Jacq.

Fl. Feb.–March. Leaves heteromorphous, often 2–4-lobed. Tubers large, a favourite food for wild hogs. In forests, uncommon.—St. Thomas (Flag Hill); St. Jan (Macumbi).

528. I. filiformis, Jacq.

Fl. Oct.–April. In thickets, often near the shore, not uncommon.—St. Croix; St. Thomas.

529. I. arenaria, Steud.

Fl. Dec.–April. Stem woody, as much as $\frac{3}{4}''$ diam. Root large, tuberous. Flowering partly precocious. On dry hills, in thickets, not uncommon.—All islands.

530. I. Quamoclit, L. (v. Sweet William).

Fl. all the year round. Near dwellings and along roads, common.— St. Croix; St. Thomas.

531. I. coccinea, L. (*I. hederæfolia*, L.).

Fl. Dec.–March. In thickets, common.—All islands.

532. I. Nil, Rth. (Bot. Mag. t. 188) (v. Morning-glory).

Fl. Oct.–March. Corolla expanded till 9 A. M. Along ditches and near dwellings, common.—All islands.

533. I. purpurea, Lam.

Fl. Oct.–Feb. Naturalized in gardens.—St. Croix; St. Thomas.

534. I. acuminata, R. S.

Fl. Nov.–March. Corolla crimson, as stated in Symb. Bot. iii, 26. Near rivulets, on trees, rare.—St. Croix (Golden Rock).

535. I. tiliacea, Chois.

St. Thomas (Schl.).

536. Jacquemontia tamnifolia, Gris.

Fl. Dec.–Feb. Seeds glabrous, greyish. In thickets, common.—All islands.

537. Convolvulus pentanthus, Jacq. (*Jacquemontia violacea*, Chois.).

Fl. Aug.–Dec. In thickets, on hills, common.—All islands.

538. C. jamaicensis, Jacq.

Fl. Dec.–Feb. In thickets, on the sandy seashore, rare.—St. Croix (Sandy Point); St. Thomas (Cowell's); Water Island.

539. C. nodiflorus, Desr. (*C albiflorus*, West) (v. Clashi-mulat).

Fl. Oct.–March. Common in thickets.—All islands.

540. C. melanostictus, Schl. (Linnæa, vi, 737).

St. Thomas (Schl.).

541. C. sagittifer, HB. Kth.

St. Thomas (Schl.).

542. Evolvulus linifolius, L.

Fl. Dec.–April. In moist localities, here and there.—All islands.

543. E. mucronatus, Sw.

Fl. Dec.–March. In marshy soil, not uncommon.—All islands.

544. E. nummularius, L.

Fl. Nov.–March. Among rocks in shady localities, not uncommon.—All islands.

545. Cuscuta americana, L. (v. Love-weed).

Fl. all the year round. In dry thickets, covering shrubs and trees, often killing them. Very common.—All islands.

(West, p. 271, mentions two species, *Convolvulus matutinus* and *C. venenatus*, as occurring in St. Croix, and refers for their description to Vahl's Symb. Bot. pars 3, as spec. nov. As, however, they are not described in any of Vahl's publications, and no specimens are in existence in Hb. Havn., I am unable to say whether they are old species or new ones.)

[Cultivated species: *Ipomœa Learii*, Annal. Fl. et Pom. 1840, p. 381, and *I. Horsfalliœ*, Hook.]

HYDROLEACEÆ.

546. Nama jamaicensis, L.

Fl. March–Aug. Among stones and rocks, a common weed.—St. Croix; St. Thomas.

BORAGINACEÆ.

547. Cordia Gerascanthus, Jacq. β) **subcanescens** (v. Rosewood, Cuppar).
Fl. Oct.   An excellent timber tree.   In forests, not very common.—
Virgin Island.

548. C. alba, R. S. (v. White Manjack).
Fl. March–Sept.   In thickets and along roads, not uncommon.—St.
Croix (eastern part of the island).

549. C. Sebestena, Jacq. α) (Bot. Mag. t. 794).   β) **rubra,** Egg. (v. Scarlet Cordia,
Fluyte boom).
Fl. all the year round.   β) leaf-ribs red; calyx scarlet as the corolla.
Both forms common in forests and planted near dwellings.—All islands.

550. C. Collococca, L. (v. Manjack).
Fl. March–April.   Precocious.   In forests, common.—All islands.

551. C. nitida, Vahl.
Fl. Jan.–Feb. and Sept.–Oct.   Flowers slightly odorous.   In forests,
not uncommon.—All islands.

552. C. lævigata, Lam.
St. Thomas (Schl.).

553. C. sulcata, DC.
Fl. June.   Leaves up to 1½' long.   In forests, not common.—Virgin
Islands; St. Croix (West, p. 275).

554. C. ulmifolia, Juss. a) **ovata,** β) **ovalis,** and γ) **lineata.**
Fl. May–Aug.   In dry thickets, common.—α) all islands ;   β) St. Thomas
(Ledru) ; γ) St. Croix (West).

555. C. cylindristachya, Sprengl.   α) **portoricensis,** Sprgl.   β) **floribunda,** Sprgl.
δ) **graveolens,** Kth.
Fl. all the year round.   On dry hills.   All three forms common.—St.
Croix ; St. Thomas.

556. C. martinicensis, R. S.
St. Croix (Griseb. Fl. p. 481).

557. C. globosa, Kth.
Fl. July–Sept.   In thickets, not uncommon.—St. Croix; St. Thomas.

558. Beurreria succulenta, Jacq. (v. Juniper).
Fl. June–Sept.   In forests and thickets, common.—All islands.

559. Rochefortia acanthophora, Gris.
Fl. June–Sept.   In thickets.—St. Croix (rare, Fair Plain, Jacob's
Peak); Virgin Islands (not uncommon).

560. Tournefortia gnaphalodes, R. Br. (v. Sea-lavender).

Fl. all the year round. On sandy shores, common.—All islands.

561. T. hirsutissima, L. (v. Chichery grape).

Fl. Sept.–April. Along roads and in thickets, especially on limestone, common.—All islands.

562. T. fœtidissima, L.

St. Croix (West, p. 270).

563. T. bicolor, Sw. β) **lævigata**, Lam.

Fl. May. Berry globose, white. Among rocks on high hills, rare.— St. Thomas (Crown, 1500′).

564. T. laurifolia, Vent.

St. Thomas (DC.).

565. T. volubilis, L.

Fl. May–Aug. Inflorescence extra-axillary, often transformed into a hollow, globose, muricate, green monstrosity, in which lives the larva of a dipterous insect. Common in thickets.—All islands.

566. T. microphylla, Desv.

Fl. May–Sept. In the same localities as the former, common.—All islands.

567. Heliotropium indicum, L.

Fl. all the year round. Along roads and in waste places, common.— All islands.

568. H. parviflorum, L. (v. Eye-bright).

Fl. all the year round. A common weed everywhere.—All islands.

569. H. curassavicum, L.

Fl. the whole year. On the sandy seashore, common.—All islands.

570. H. fruticosum, L.

Fl. all the year round. Up to 6′ high. On dry hills.—St. Croix (common in the eastern part); Virgin Islands (not uncommon).

[Cultivated species: *H. peruvianum*, L. (v. Heliotrope.)]

POLEMONIACEÆ.

[Cultivated in gardens: *Phlox Drummondii*, Hook.]

SOLANACEÆ.

571. Brunfelsia americana, Sw. α) and β) **pubescens** (v. Rain-tree).

Fl. May–Dec. Flowers odorous before rain. In thickets and woods, common.—Virgin Islands (cultivated in gardens in St. Croix).

572. Datura Metel, L. (v. Fire-weed).

Fl. all the year round. Flowers nocturnal. Along roads and in waste places, naturalized everywhere.—All islands.

573. D. fastuosa, L.

Fl. all the year round. Naturalized in gardens and near dwellings.—All islands.

574. D. Tatula, L.

Fl. May–Dec. Along roads, naturalized, but rare.—St. Croix (Hope).

575. D. Stramonium, L. (v. Fire-weed).

Fl. Sept.–Feb. Naturalized in waste places, common.—All islands.

576. Nicotiana Tabacum, L.

Fl. May–Nov. Used as a medicine, but not for smoking. Naturalized near dwellings.—All islands.

577. Physalis peruviana, L.

Fl. May–Nov. In fields, uncommon.—St. Thomas (Rapoon).

578. P. pubescens, L.

Fl. March–May. In shady valleys, uncommon.—St. Croix (Crequis); St. Thomas.

579. P. Linkiana, Ns.

Fl. Dec. In cultivated fields, not uncommon.—St. Thomas.

580. P. angulata, L.

Fl. Sept.–Jan. Stamens of unequal length; anthers successively dehiscent. Along roads and ditches, common.—All islands.

581. Capsicum dulce, Hort. (DC. Prodr. xiii, i, 428) (v. Sweet Pepper).

Fl. March–July. Berry oblong. Naturalized in gardens.—St. Croix; St. Thomas.

582. C. frutescens, L. (v. Bird Pepper).

Fl. Aug.–Dec. Used as a condiment. Here and there in forests and cultivated.—St. Croix; St. Thomas.

583. C. baccatum, L. (v. Small Pepper).

Fl. Aug.–Jan. In forests and near dwellings, not uncommon.—All islands.

584. C. annuum, L. (v. Pepper).

Fl. all the year round. Fruit universally used as a condiment. Cultivated and naturalized everywhere.—All islands.

585. Lycopersicum cerasiforme, Dun. (Solan. p. 113) (v. Small Trovo).

Fl. May–Sept. Berry globose, small, yellow. Not uncommon near dwellings (perhaps only naturalized). Used as a vegetable.—St. Croix; St. Thomas.

586. L. esculentum, Mill. (v. Tomato, Trovo).

Fl. all the year round. Berry used as a vegetable. Cultivated and naturalized everywhere.—All islands.

587. Solanum nodiflorum, Jacq. *a*) and *β*) oleraceum, Dun. (v. Lumbush).

Fl. May–Dec. Stem often prickly. In fields and in waste places, common.—All islands.

588. S. verbascifolium, L. (v. Turkey-berry).

Fl. June–Oct. In waste places, not uncommon.—Virgin Islands; St. Croix (West, p. 274).

589. S. racemosum, L. (v. Canker-berry).

Fl. all the year round. Proterandrous. In waste places, very common.—All islands.

590. S. igneum, L. (v. Canker-berry).

Fl. all the year round. Habitat of the preceding. Very common.—All islands.

591. S. bahamense, L. (*S. persicæfolium,* Dun.)

Fl. Jan.–Aug. Along coasts, not uncommon.—Virgin Islands.

592. S. lanceifolium, Jacq.

Not seen flowering. Leaves and stem very prickly. In forests, rare.—St. Jan (King's Hill, 1000′).

593. S. torvum, Sw. (v. Plate-bush).

Fl. all the year round. A shrub or small tree. In forests and near dwellings, common.—All islands.

594. S. inclusum, Gris., var. **albiflorum,** Egg.

Fl. all the year round. Corolla white, $\frac{3}{4}''$–1″ diam. Stigma 3–5-branched, stellate. Berry globose, somewhat depressed, hirsute, orange-coloured, 1″ diam. The excrescent calyx prickly. In dry thickets, not uncommon.—Virgin Islands.

595. S. aculeatissimum, Jacq.

Fl. April–May. Naturalized by mules from Montevideo.—St. Croix (Frederiksted).

596. S. mammosum, L.

St. Croix (West, p. 275).

597. S. polygamum, Vahl (v. Kakkerlakka-berry).

Fl. all the year round. In dry thickets, common.—Virgin Islands.
(In DC. Prodr. xiii, i, 197, it is stated that this species has been found in St. Croix by Wydler, which, however, appears doubtful to me. West, p. 275, only gives St. Jan as habitat, yet Vahl in his Symb. Bot. iii, 39, and after him probably Griseb. Fl. p. 443, refer to West as the authority for St. Croix as habitat.)

598. Cestrum laurifolium, L'Her.

Fl. Jan.–April. Petiole black; berry dark purple.· In forests, not uncommon.—All islands.

599. C. diurnum, L.

Fl. Feb.–June. In forests, uncommon.—Virgin Islands; St. Croix (West, p. 276).

600. C. nocturnum, L.

Fl. March. In forests, rare.—St. Jan (Rogiers, Joshee Gut).

[Cultivated species: *Datura suaveolens,* HBK.; *Petunia nyctaginiflora,* Juss., and *P. violacea,* Lindl.; *Solanum Seaforthianum,* Andr., *S. tuberosum,* L. (v. Irish potato), and *S. Melongena,* L. (v. Egg-plant, Beranger).]

SCROPHULARIACEÆ.

601. Scoparia dulcis, L.

Fl. all the year round. A common weed along roads and in moist localities.—All islands.

602. Capraria biflora, L. a) and β) **pilosa** (v. Goat-weed).

Fl. all the year round. Leaves used for tea. Both forms along roads, common. a) in moist, β) in dry localities.—All islands.

603. Herpestis stricta, Schrad.

St. Thomas (Benth.).

604. H. chamædryoides, Kth.

Fl. Dec.–March. Pedicel bearing two bracteolæ at the base. The two innermost calyx-lobes setaceous. In moist localities, rare.—St. Croix (Spring-gut).

605. H. Monniera, Kth.

Fl. all the year round. Along rivulets and on the margins of lagoons, common.—All islands.

606. Vandellia diffusa, L.

St. Croix (Ryan in Hb. Havn., Vahl's Eclogue, ii, 47) (Montserrat, Ryan in Hb. Havn., "*vulgaris*").

[Cultivated species: *Maurandia Barclayana*, Lindl. (v. Fairy Ivy), and *Russelia juncea*, Zucc. (v. Madeira Plant).]

BIGNONIACEÆ.

607. Crescentia Cujete, L. (v. Calabash-tree).

Fl. all the year round. Leaves deciduous in Dec. The fruit is used for vessels. Near dwellings and in forests, common.—All islands.

608. C. cucurbitina, L. (v. Black Calabash).

Fl. March–Nov. Wood used for boat-building. In dense forests near rivulets, not uncommon.—All islands.

609. Catalpa longisiliqua, Cham.

St. Thomas (Gris. Fl. 446).

610. Tecoma Berterii, DC.

Fl. March–July. Leaves deciduous Feb.–April. In dry thickets, common.—Virgin Islands.

611. T. leucoxylon, Mart. (v. White Cedar).

Fl. March–April, precocious, and later coëtanous in Sept.–Oct. Wood used for building boats. In forests and on dry hills, common.—All islands.

612. T. stans, Juss. (v. Yellow Cedar).

Fl. all the year round. Anthers pilose beneath. In thickets, common; often gregarious, especially in St. Croix.—All islands.

613. Bignonia æquinoctialis, L.

Fl. April–Sept. Anthers pilose or glabrous (hence Vahl's distinction on this account between his *B. spectabilis* (Symb. Bot. iii, p. 80) and this species not justified). Here and there in marshy forests.—St. Thomas (Northside Bay, Sta. Maria); St. Croix (Salomon's estate, West, p. 294).

614. B. unguis, L. (v. Cat-claw).

Fl. April–May, precocious, later again coëtanous in Nov. Stem 1½″ diam., showing the irregular structure peculiar to all climbing *Bignoniaceæ*. Fruit as much as 26″ long. In forests, not uncommon.—All islands.

615. Distictis lactiflora, DC. (Prodr. ix, 191) (*Bignonia*, Vahl).

Fl. all the year round. On fences and in dry thickets, here and there.—St. Croix (Cotton Grove, Southgate Farm) (cultivated in St. Thomas). •

[Cultivated species: *Tecoma capensis*, Lindl.]

ACANTHACEÆ.

616. Ruellia tuberosa, L. (v. Christmas-pride).

Fl. all the year round; most abundantly towards Christmas. Along roads and ditches, common.—All islands.

617. R. strepens, L.

St. Croix (Isert sec. DC. Prodr. xi, 121).

618. Stemonacanthus coccineus, Gris.

Fl. Jan.–April. Cleistogamous flowers in July; also an intermediate form between cleistogamous and normal flowers. In shady forests, rare.—St. Croix (Caledonia, Wills Bay); St. Jan (Bordeaux Hills); St. Thomas (Wydl. sec. DC. Prodr. xi, 217).

619. Blechnum Brownei, Juss. (v. Penguin Balsam).

Fl. Dec.–April. Used against cough. In pastures and along ditches, common.—All islands.

620. Barleria lupulina, Lindl. (Bot. Reg. t. 1483).

Fl. Dec.–April. Naturalized near dwellings and in gardens.—St. Thomas; St. Jan.

621. Thyrsacanthus nitidus, Ns.

St. Croix (v. Rohr sec. Symb. Bot. ii, 5, and Isert sec. DC. Prodr. xi, 327); St. Thomas (Nees).

622. Dianthera pectoralis, Murr. (v. Garden Balsam).

Fl. Dec.–March. Used against coughs. Naturalized near dwellings and in gardens.—All islands.

623. D. sessilis, Gris. (*Justicia pauciflora*, Vahl in Eclog. Am. i, 2).

Fl. June–July. Flowers often cleistogamous. Rhizome perennial. In thickets, here and there.—St. Croix (Salt River); St. Thomas.

624. Justicia carthagenensis, Jacq.

Fl. Dec.–March. Along ditches and in forests.—All islands.

625. J. reflexiflora, Rich. (Vahl's Enum. Plant, i, 157), var. glandulosa, Egg.

Fl. all the year round. Bracts densely glanduliferous. Seeds globose, brown. Procumbent among bushes.—St. Croix (rare, Fair Plain); St. Thomas; Buck Island (not uncommon).

626. J. periplocæfolia, Jacq.

St. Thomas (Schl.).

627. Beloperone nemorosa, Nees.

Fl. Jan.–March. Calyx one-sixth of the length of the corolla. In forests, rare.—St. Croix (Caledonia, Ham's Bluff Valley).

628. Crossandra infundibuliformis, Nees.

Fl. March–June. Naturalized in gardens.—St. Croix.

629. Stenandrium rupestre, Ns. (DC. Prodr. xi, 283) (*Ruellia?*, Sw. Fl. Ind. Occ. p. 1071; Plum. Icon. ed. Burm. t. 75, as *Gerardia*). c) glabrous, β) pilose.

Fl. Dec.–May, cleistogamous. Normal flowers June–Aug. Corolla expanded till 9 A. M. Rhizome perennial; roots fusiform, tuberous. Gregarious on the ground in forests, rare.—a) St. Thomas (Flag Hill, 700'–900'); β) St. Jan (Baas Gut).

630. Anthacanthus spinosus, Nees.

Fl. all the year round. Flowers heterostylous. On rocks and in forests, common, especially in St. Croix.—All islands.

631. A. jamaicensis, Gris.

Fl. June–July. Corolla-lobes glandular inside. On limestone, rare.— St. Croix, in stony ground.

632. A. microphyllus, Ns.

Fl. May–Aug. In forests, here and there.—All islands.

633. Dicliptera adsurgens, Juss.

Fl. Jan.–Feb., cleistogamous; normal, March–April. In thickets and near ditches.—St. Croix (common); St. Jan (less common).

634. Thunbergia volubilis, Pers.

Fl. all the year round. Naturalized along ditches and rivulets.—St. Croix (Caledonia, Mt. Stewart); St. Thomas (Tutu).

[Cultivated species: *Graptophyllum hortense*, Nees, *Justicia bicolor*, Andr., *Thunbergia alata*, Boj., *Th. fragrans*, Roxb., and *Sesamum orientale*, L. (v. Benye).]

GESNERIACEÆ.

635. Martynia diandra, Glox. (v. Cocks).

Fl. Sept.–Dec. Three rudimentary filaments; 1'–3' high. Along roads and in waste places, not uncommon.—St. Croix; St. Thomas.

LABIATÆ.

636. Ocimum Basilicum, L.

Fl. May–Aug. Naturalized in gardens.—All islands.

637. O. micranthum, W. (v. Passia Balsam).

Fl. Aug.–Nov. Corolla expanded during the morning. Used against coughs. Along ditches and in pastures, gregarious.—All islands.

638. Coleus amboinicus, L. (v. East India Thyme).

Fl. April–May. Naturalized in dry localities, gregarious. — All islands.

639. Hyptis capitata, Jacq. (v. Wild Hops).

Fl. Nov.-March. Along rivulets, common.—St. Croix; St. Thomas.

640. H. suaveolens, Poit.

Fl. Oct.-Feb. 3'-4' high. In dry localities, common.—St. Croix; St. Thomas.

641. H. pectinata, Poit. (v. French Tea).

Fl. Nov.-April. As much as 8' high. In dry localities, not uncommon.—All islands.

642. H. verticillata, Jacq.

St. Thomas (Gris. Fl. p. 489).

643. Salvia occidentalis, Sw.

Fl. Dec.-March. Rhizome thick. Along roads, common.—All islands.

644. S. tenella, Sw.

St. Thomas (Gris. Fl. p. 490; Schl.).

645. S. serotina, L.

Fl. Sept.-April. Leaves very bitter. Corolla white. In dry localities, gregarious, common.—All islands.

646. S. coccinea, L. a) and β) ciliata, Benth.

Fl. all the year round. Along ditches and roads, common.—All islands.

647. Leonurus sibiricus, L.

Fl. all the year round. Corolla sometimes white. A common weed in fields and along roads.—All islands.

648. Leucas martinicensis, R. Br.

Fl. March-Nov. A weed, common in gardens and along roads.—St. Croix.

649. Leonotis nepetæfolia, R. Br. (v. Hollow Stock).

Fl. all the year round. Corolla sometimes white. Gregarious, a very common weed everywhere.—All islands.

650. Mentha aquatica, L. (v. Mint).

Not seen flowering. Naturalized along rivulets, gregarious.—St. Croix (Caledonia).

[Cultivated species: *Rosmarinus officinalis*, L. (v. Rosemary), *Thymus vulgaris*, L. (v. Thyme), and *Origanum Majorana*, L. (v. Sweet Marjoram Tea).]

VERBENACEÆ.

651. Priva echinata, Juss.

Fl. all the year round.　Corolla expanded till 10 A. M.　A common weed along roads and in gardens.—All islands.

652. Bouchea Ehrenbergii, Cham.

Fl. Dec.–May.　Gregarious along roads and in dry localities, common.—St. Croix; St. Thomas.

653. Stachytarpha jamaicensis, V. (v. Vervain).

Fl. all the year round.　Flower expanded till noon.　Pollen 3–4-branched, stellate.　Leaves used against fever.　Very common along roads and ditches.—All islands.

654. S. strigosa, Vahl.

St. Thomas (Ehrenb. sec. DC. Prodr. xi, 564; Gris. Fl. p. 494).

655. Lippia nodiflora, Rich.

Fl. all the year round.　Gregarious in moist localities, not uncommon.—St. Croix (La Reine, Fair Plain).

656. Lantana Camara, L. (v. Sage).

Fl. all the year round.　Berry considered to be poisonous.　On dry hills, very common.—All islands.

657. L. polyacantha, Schauer (DC. Prodr. xi, 597) (*L. scabrida*, Ait.).

Fl. all the year round.　In dry localities, here and there.—St. Croix (St. George); St. Thomas (Solberg).

658. L. involucrata, L.

Fl. all the year round.　Corolla and berry violet.　In thickets, common, especially on limestone.—All islands.

659. L. reticulata, Pers.

Fl. all the year round.　On limestone, rare.—St. Croix, in stony ground (King's Hill).

660. Citharexylum quadrangulare, Jacq. (v. Fiddlewood, Susanna).

Fl. July–Sept.　In forests, not uncommon.—St. Croix; St. Thomas.

661. C. cinereum, L. (v. Susanna).

Fl. July–Dec.　Leaves of both these species becoming red in Feb., and dropping off at the same time that the new ones make their appearance.　On young radical shoots the leaves are linear and deeply serrate. The wood is quite useless, even for firewood.　In dry thickets and forests, common, often gregarious.—All islands.

662. C. villosum, Jacq. (Icon. Var. t. 118).

St. Thomas (Schlecht., Bertero, Duchass. sec. Gris. Syst. Unt.).

663. Duranta Plumieri, Jacq.

Fl. May–Dec.   Along roads and in thickets, common.—All islands.

664. Callicarpa reticulata, Sw.

St. Croix (West, p. 269).

665. Ægiphila martinicensis, Jacq.

Fl. Aug.–Jan.   Flowers often heterostylous.   In forests, common.—St. Croix.

666. Clerodendron aculeatum, L. (v. Chuc-chuc).

Fl. all the year round.   Common on dry hills and in marshy soil.—All islands.

667. C. fragrans, W.

Fl. all the year round.   Long creeping rhizome.   Gregarious on high hills in shady places, naturalized.—St. Thomas (Dorothea, Liliendal).

668. Petitia domingensis, Jacq.   a).

Fl. May–Sept.   Leaves often ternate.   Drupe commonly 4-loculate. A tree up to 50' high.   In forests, not uncommon.—St. Croix (Caledonia, Punch, Wills Bay).

669. Vitex divaricata, Sw.

Fl. May–July.   Filaments glandular-pilose.   A low tree, here and there in forests.—St. Croix (Caledonia, Wills Bay); St. Thomas (Crown); St. Jan (Cinnamon Bay).

670. Avicennia nitida, Jacq.

Fl. all the year round.   Upper surface of leaves always covered with small salt crystals.   Along the seashore and lagoons, common.—All islands.

671. A. tomentosa, Jacq.

St. Croix (West, p. 269); St. Thomas (Schl.).

[Cultivated species: *Verbena chamædrifolia*, Juss., in several varieties, *Petræa volubilis*, Jacq. (v. Wreath-plant), *Aloysia citriodora*, Ortega (v. Lemon-scented Verbena), *Vitex Agnus-castus*, L. (v. Wild Black Pepper), and *Holmskjoldia sanguinea*, Retz.]

MYOPORACEÆ.

672. Bontia daphnoides, L. (v. White Alling).

Fl. all the year round.   On sandy shores.—St. Croix (rare, Turner's Hole); Virgin Islands (not uncommon).

PLANTAGINACEÆ.

673. Plantago major, L. β) tropica (v. English Plantain).

Fl. Jan.–March. Proterogynous. Leaves used against inflammation of the eyes.

PLUMBAGINACEÆ.

674. Plumbago scandens, Thunb. (v. Blister-leaf).

Fl. all the year round. Leaves used as blisters. In thickets and forests, common.—All islands.

[Cultivated species: *P. capensis*, Thunb.]

PHYTOLACCACEÆ.

675. Suriana maritima, L.

Fl. June–Dec. Stamens mostly 10. Filaments pilose. On sandy shores, not uncommon.—All islands.

676. Microtea debilis, Sw.

Fl. July–Sept. In shady places, rare.—St. Croix (Spring Garden, Wills Bay).

677. Rivina lævis, L. (v. Snake-bush, Stark mahart). α) and β) **pubescens.**

Fl. all the year round. A common weed everywhere, both forms.—All islands.

678. R. octandra, L.

Fl. Feb.–Aug. Pedicel and calyx becoming reddish-brown as well as the fruit. Stamens in two whorls, mostly 12. In thickets and forests, common.—All islands.

679. Petiveria alliacea, L. (v. Gully-root).

Fl. all the year round. A very common weed everywhere.—All islands.

CHENOPODIACEÆ.

680. Chenopodium ambrosioides, L.

Fl. March. In waste places and on walls, here and there.—St. Croix (Fredriksted); St. Jan (Cruz Bay).

681. Ch. murale, L.

Fl. Jan.–May. On walls, uncommon, naturalized.—St. Croix; St. Thomas.

682. Obione cristata, Moq. (DC. Prodr. xiii, ii, p. 110).

Fl. March–Aug. On sandy shores, uncommon.—St. Thomas (Water Bay); St. Jan; St. Croix (Schl.).

683. Boussingaultia baselloides, Kth. (Bot. Mag. t. 3620).

Fl. all the year round. Naturalized in gardens and cultivated.—St. Croix; St. Thomas.

684. Batis maritima, L.

Fl. all the year round. Gregarious along the coast of lagoons, common.—St. Croix; St. Thomas.

[Cultivated species: *Beta vulgaris*, L. (v. Red Beet).]

AMARANTACEÆ.

685. Celosia argentea, L. (*C. margaritacea*, L.).

Fl. all the year round. Naturalized around dwellings.—St. Thomas; St. Croix (West, p. 277).

686. C. nitida, Vahl.

Fl. all the year round. In forests and thickets, not uncommon.—St. Croix; St. Thomas.

687. Chamissoa altissima, Kth.

Fl. Dec.–March. In forests, here and there.—St. Croix (Lebanon Hill); St. Thomas (Signal Hill).

688. Achyranthes aspera, L.   α) argentea, Lam.   β) obtusifolia, Lam.

Fl. Dec.–March. In thickets and on waste places, common.—All islands.

689. Gomphrena globosa, L. (v. Bachelor's Button).

Fl. all the year round. Naturalized in gardens and near dwellings.—All islands.

690. Iresine elatior, Rich.

Fl. Sept.–March. Uppermost leaves always alternate. In thickets, common.—All islands.

691. Philoxerus vermiculatus, R. Br. (v. Bay-flower).

Fl. all the year round. Along the coast, very common, gregarious.—All islands.

692. Alternanthera polygonoides, R. Br.   a).

Fl. all the year round. In sandy places, common.—All islands.

693. A. ficoidea, R. Br.

Fl. all the year round. In moist localities, uncommon.—St. Thomas (Haulover).

694. A. Achyrantha, R. Br.

Fl. March–Aug. Among rocks and stones, here and there.—St. Croix, St. Thomas (Schl.).

695. Amblogyne polygonoides, Raf.

Fl. all the year round: ♂ flowers very few. In sandy places near the coast, common.—St. Croix; St. Thomas.

696. Scleropus amarantoides, Schrad.

Fl. all the year round. Leaves often discoloured with white cross-stripes. In sandy localities, common.—All islands.

697. Euxolus caudatus, Moq.

Fl. all the year round. In waste places, common.—All islands.

698. E. oleraceus, Moq. (v. Lumbo).

Fl. all the year round. Near dwellings, common.—All islands.

699. Amarantus spinosus, L.

Fl. Jan.–April. Near rivulets and ditches, uncommon.—St. Croix; St. Thomas.

700. A. tristis, L.

St. Thomas (Wydler sec. DC. Prodr. xiii, ii, 260).

701. A. paniculatus, L. (v. Bower).

Fl. all the year round. A troublesome weed on account of its long tap-root. Common everywhere.—All islands.

NYCTAGINACEÆ.

702. Mirabilis Jalapa, L. (v. Four-o'clock).

Fl. all the year round. Flower expanded from 4 P. M., purple, yellow, or pink. Around dwellings, common.—All islands.

703. Boerhaavia erecta, L.

Fl. Dec.–Feb. Along ditches and in pastures, uncommon.—St. Croix (Mt. Stewart).

704. B. paniculata, Rich. (v. Batta-batta).

Fl. all the year round. Calyx often transformed into a hollow monstrosity by the larva of a wasp. A very common weed.—All islands.

705. Pisonia aculeata, L.

Fl. Feb.–April. In forests, common.—St. Croix; St. Thomas.

706. P. subcordata, Sw. (v. Mampoo, Loblolly).

Fl. April–June. Leaves partly deciduous. Wood useless for timber and fuel. Along coasts, common, growing to a large tree.—All islands.

707. P. inermis, Jacq.

Fl. April–May. Leaves on the young branches whorled. In forests, common.—All islands.

[Cultivated species: *Bougainvillea spectabilis,* Willd.]

POLYGONACEÆ.

708. Ccccoloba uvifera, Jacq. (v. Sea-grape).

Fl. July–Dec. Wood hard, dark purple, used for ship-building. On the sandy seashore, common. Sometimes in the interior as high up as 1200′.—All islands.

709. C. leoganensis, Jacq.

Fl. May–July. Flowers in fascicles of 3–4, of which, however, one only bears fruit. Drupe oval, violet, 4‴ long. On sandy shores, rare.—St. Croix (Sandy Point).

710. C. rugosa, Desf. (DC. Prodr. xiv, 152; Bot. Mag. t. 4536).

St. Thomas (DC. Prodr. l. c.).

711. C. laurifolia, Jacq. (Hort. Schœnbr. iii, p. 9, t. 267).

Fl. March–July. Leaves deciduous April to May. Fruit purplish, pointed at both ends. In thickets, here and there.—St. Croix (Sandy Point, Hard Labour).

712. C. diversifolia, Jacq.

Fl. May–July. 6′–8′ high. Along the coast, uncommon.—St. Croix (La Vallée, Claremont).

713. C. obtusifolia, Jacq.

St. Croix (West, p. 281).

714. C. punctata, Jacq. α) **Jacquinii,** β) **barbadensis,** Jacq., δ) **parvifolia** (v. Red wood, Roehout), γ) **microstachya,** W.

Fl. Aug.–Dec. α) leaves as much as 1⅜′ long. A shrub or low tree. δ) and γ) common; α) and β) uncommon.—All islands.

715. C. nivea, Jacq.

Fl. June–Sept. Flowers delicately odorous. Fruit white when ripe. In forests, not uncommon.—All islands.

(*C. Klotzschiana*, Meissn., and *C. Kunthiana*, Meissn. (DC. Prodr. xiv, 155 and 166), are said to have been found in St. Thomas, but they are both very doubtful species, founded on single specimens, and have therefore been here omitted.)

[Cultivated species: *Antigonon cordatum*, Mart. & Galeotti (v. Mexican Wreath-plant), and *Rumex vesicarius*, L.]

LAURACEÆ.

716. Cinnamomum zeilanicum, Bl.

Fl. April–May. Naturalized in a few places in shady valleys.—St. Croix (Crequis).

717. Phœbe antillana, Meissn. (DC. Prodr. xv, i, p. 31). γ) **cubensis.**

St. Croix (West in IIb. Petrop. sec. DC. l. c.).

(*Ph. montana*, Gris., said by Meissn. (DC. Prodr. l. c. p. 236) to be synonymous with *Laurus longifolia*, Vahl, mentioned by West, p. 262, as a new species from St. Croix, ought perhaps to be added to this list; but as the specimens seen by me in IIb. Havn. as *Laurus longifolia*, Vahl, do not agree with Grisebach's, I prefer to omit the species here, as being doubtful.)

718. Persea gratissima, Gaertn. (v. Alligator Pear).

Fl. March–May. Stamens, 9 perfect, 3 less perfect and sterile, 6 rudimentary. The fruit is a favourite vegetable. In gardens.—All islands.

719. Hufelandia pendula, Ns. (*H. Thomæa*, Nees).

St. Thomas (sec. DC. Prodr. l. c. p. 65, IIb. Kunth!).

720. Acrodiclidium salicifolium, Gris.

Fl. May–Aug. In forests, here and there.—St. Croix (Wills Bay, Spring-gut).

721. Nectandra coriacea, Gris.

Fl. May–Aug. In forests, rare.—St. Thomas (Soldier Bay); St. Jan (IIb. Havn.).

722. N. membranacea, Gris.

Fl. June. In dense forests, uncommon.—St. Croix (Wills Bay); St. Thomas (Signal Hill).

723. N. antillana, Meissn. (DC. Prodr. l. c. 153) (*N. leucantha*, Gris.).

Fl. May–June. In forests, not uncommon. Fragrant.—All islands.

724. Oreodaphne leucoxylon, Nees.

Fl. July. In dense forests on high hills, uncommon.—St. Thomas (Signal Hill) (Montserrat, Ryan in IIb. Havn.).

725. Cassyta americana, L.

Fl. March–April. Inflorescence often branched. On Manchineel and Acacia trees along the seashore, here and there.—St. Croix (Cotton Grove); St. Thomas (Water Bay); Vieques (IIb. Havn.).

THYMELÆACEÆ.

726. Daphnopsis caribæa, Gris.

Fl. July and Dec.–March. In forests, not uncommon.—St. Thomas (Flag Hill, Signal Hill).

EUPHORBIACEÆ.

727. Buxus Vahlii, Baill. (DC. Prodr. xvi, i, p. 16) (*Tricera lævigata*, Sw., var. *Sanctæ-Crucis*, Eggers in Fl. St. Crucis, p. 111).

Fl. June–Oct. On limestone, rare.—St. Croix (Stony Ground).

728. Savia sessiliflora, W. (Spec. Plant. iv, p. 771).

Fl. June–Dec. In thickets on dry hills, not uncommon.—All islands.

729. Phyllanthus acuminatus, Vahl (Symb. Bot. ii, 95).

St. Thomas (Herb. DC. sec. DC. Prodr. xv, ii, 381). Vahl, however, gives only Cayenne (Rohr) as habitat.

730. Ph. Niruri, L. (v. Creole Chinine).

Fl. all the year round. Very common in gardens and along roads.—All islands.

731. Ph. distichus, Müll. (DC. Prodr. l. c. 413) (*Cicca*, L.) (v. Gooseberry).

Fl. June–Sept. Fruit used for preserves. Naturalized near dwellings.—All islands.

732. Ph. nobilis, Müll. (l. c. 415). η) **Antillana** (*Cicca*, Juss.) (v. Gongora-hout).

Fl. July, and afterwards precocious in Dec.–Jan. In forests, not uncommon.—All islands.

733. Ph. falcatus, Sw. (v. Boxwood).

Fl. all the year round. In marshy soil, not uncommon.—Vieques.

734. Securinega acidothamnus, Müll. (l. c. 451) (*Flüggea*, Gris.).

Fl. May–June. In thickets, not uncommon.—St. Croix (eastern part of the island).

(I have adopted Müller's generic name, *Flüggea* being an older name for a genus of *Ophiopogoneæ* established by L. C. Richard.)

735. Drypetes lævigata, Gris. ined. (*Excæcaria polyandra*, Gris. Cat. Pl. Cub. p. 20, & Diagnos. neuer Euphorb. p. 180).

Fl. Sept. ♂. I have not found the female flower nor fruit, and am therefore not able to supply the deficiency in this respect in Grisebach's Diagnosis.—St. Croix (Fair Plain); St. Jan (Cinnamon Bay).

736. D. glauca, Vahl.

St. Croix (Hb. Havn. Ryan, Rohr; "Hollow berry of Bugby Hole") (Montserrat, Ryan in Hb. Havn.).

737. Croton astroites, Ait. (v. White Marán).

Fl. Dec.–July. Style 16-branched. In dry thickets, very common.—All islands.

738. C. betulinus, Vahl (Symb. Bot. ii, p. 98).

Fl. all the year round. A low shrub, brownish. Common in thickets.—All islands.

739. C. flavens, L. (v. Marón).

Fl. all the year round. Gregarious on dry hills, also as secondary growth; very common, and a troublesome shrubby weed.—All islands.

740. C. discolor, Willd. (Spec. Plant. iv, 352) (*C. balsamifer,* L.).

Fl. all the year round. Along roads in dry localities, common.—St. Croix (eastern part of the island); St. Thomas (IIb. Thunb. sec. DC. Prodr. l. c. p. 615).

741. C. oval'folius, West.

Fl. all the year round. Along roads and in waste places, very common.—All islands.

742. C. lobatus, L.

Fl. March–Dec. In the same places as the preceding, very common.—All islands.

743. C. humilis, L.

St. Thomas (Bertero sec. DC. Prodr. l. c. 670).

(An arboreous as yet undetermined *Crotonea,* not found in blossom, occurs in a few specimens on Flag Hill in St. Thomas.)

744. Aleurites Moluccana, Willd. (Spec. Plant. iv, 5£0) (*A. triloba,* Forst.) (v. Walnut).

Fl. all the year round. Naturalized near dwellings and in gardens.—St. Croix; St. Thomas.

745. Ricinella pedunculosa, Müll. (Linnæa, xxxiv, 153) (*Adelia Ricinella,* L.).

Fl. March–May, precocious. Always very spiny. In dry thickets, not uncommon.—All islands.

746. Argyrothamnia fasciculata, Müll. (Linnæa, l. c. 146) (*Ditaxis,* Schl.).

Fl. Jan.–May and Sept. In thickets, not uncommon.—All islands.

747. A. candicans, Müll. (DC. Prodr. l. c. 741) (*Argythamnia,* Sw.).

Fl. Sept.–April. Capsule dark blue; seeds verrucose. In thickets, common.—All islands.

748. Acalypha chamædrifolia, Müll. (l. c. 879). β) **genuina** (*A. reptans,* Sw.), γ) **brevipes.**

Fl. all the year round; female flowers developing gradually. Bracts persistent after dissemination. On rocks and in crevices, not uncommon.—St. Croix (β); St. Thomas (γ).

749. Tragia volubilis, L. (v. Nettle, Bran-nettle).

Fl. Feb., Sept. Male flowers often transformed into a globose mon-strosity. The plant is believed by the negroes to give them luck in marketing. In thickets and along roads, common.—All islands.

750. Ricinus communis, L. a) (v. Castor-oil tree).

Fl. all the year round. Seeds used for pressing castor-oil. Naturalized on waste places, common.—All islands.

751. Manihot utilissima, Pohl (Plant. Bras. i, 32) (v. Cassava).

Fl. March–May. Root used for manufacturing starch and flour, which is made up into flat, thin cakes (bambam). Naturalized and culti-vated.—All islands.

752. Jatropha Curcas, L. (v. French Physic-nut, Skituetchi).

Fl. all the year round. Seeds very drastic. A low tree, often planted on graves. Naturalized near dwellings, common.—All islands.

753. J. gossypiifolia, L. (v. Physic-nut). a) **staphysagriæfolia,** β) **elegans.**

Fl. all the year round. The whole plant has a disagreeable smell. Suffrutescent, 1′–4′ high. ' A troublesome weed near dwellings and in fields. Very common everywhere.—All islands.

754. J. multifida, L. (v. Coral-bush).

Fl. all the year round. Naturalized in gardens.—St. Croix; St. Thomas.

755. Sebastiania lucida, Müll. (DC. Prodr. l. c. 1181) (*Excœcaria*, Sw.).

Fl. Feb.–June. A shrub or low tree, 5′–20′ high. In thickets and forests, common.—All islands.

756. Hippomane Mancinella, L. (v. Manchineel-tree).

Fl. precocious, Feb.–April, coëtanous, May–June. Wood affording excellent timber, but very little used on account of the caustic milky juice. On sandy shores, often gregarious, sometimes in the interior of the islands on hills.—St. Croix (common); Virgin Islands (uncommon).

757. Excœcaria Laurocerasus, Müll. (l. c. 1202). γ) **laurifolia.**

Not seen flowering. A high tree; bark smooth, white. In dense forests, rare.—St. Jan (Cinnamon Bay).

758. Hura crepitans, L. (v. Sandbox-tree).

Fl. Sept. Leaves deciduous in Jan.–April. Seeds drastic. A high tree with horizontal branches and prickly stem. In forests and near dwellings, common.—All islands.

759. Dalechampia scandens, L.

Fl. Feb.–June. Male inflorescence bearing at the base two resinous corpuscula, deciduous together with the male flowers. Baillon considers them to be sterile bracts; Müller takes them for monstrous anthers. Central female flower pedicellate. In thickets, common.—All islands.

760. Euphorbia buxifolia, Lam.

Fl. all the year round. On the sandy shore, common.—All islands.

761. E. articulata, Burm.

Fl. all the year round. Along the seacoast, common.—All islands.

762. E. pilulifera, L.

Fl. all the year round. In waste places and along roads, very common.—All islands.

763. E. hypericifolia, L. a) and β) **hyssopifolia, L.**

Fl. all the year round. Leaves distichous. Used against dysentery. Same places as the preceding. A common weed.—All islands.

764. E. thymifolia, Burm.

Fl. all the year round. The whole plant reddish. Leaves folding together during night and in rainy weather. Among stones and along roads, very common.—All islands.

765. E. prostrata, Ait.

Fl. the whole year. Together with the preceding, common.—All islands.

766. E. petiolaris, Sims (Bot. Mag. t. 883) (v. Manchineel).

Fl. the whole year. Partly precocious in the spring. On dry hills and in thickets.—Virgin Islands (common); St. Croix (West, p. 288?).

(West's *E. cotinifolia*, said to occur in St. Croix, is evidently meant for this species. I doubt, however, the correctness of the habitat, and am of opinion that it is a mistake for St. Thomas, where the species is exceedingly common.)

767. E. geniculata, Ortega (Decad. p. 16; DC. Prodr. xv, ii, 72). (*E. prunifolia*, Jacq. Hort. Schœnbr. iii, t. 277, a form with larger, serrate leaves.)

Fl. Dec.–March. In forests and near dwellings, not uncommon, often gregarious.—St. Croix (Government House); St. Thomas (Signal Hill).

768. E. heterophylla, L. β) **cyathophora, Jacq.**

Fl. all the year round. Gregarious in dry places, common.—All islands.

769. E. neriifolia, L. (DC. Plant. Grasses, i, t. 46).

Fl. March–June. A large tree, stem 2'–3' diam. Naturalized near dwellings, common.—All islands.

770. Pedilanthus tithymaloides, Poit. *a*), *β*) **padifolius,** Poit., and *γ*) **angusti-folius,** Poit.

Fl. all the year round.   In thickets and gardens, uncommon.—All islands.

All *Euphorbiaceæ* are proterogynous.

[Cultivated species: *Jatropha panduræfolia,* Andr., *Codiæum variegatum,* Müll. *a*) *pictum, Euphorbia pulcherrima,* W., *E. splendens,* Boj., and *E. antiquorum,* L.]

URTICACEÆ.

771. Celtis trinervia, Lam.

Fl. June–Dec.   In forests and thickets, not uncommon.—All islands.

772. C. aculeata, Sw. *a*) and *β*) **serrata.**

Fl. March–Sept.   Proterogynous.   Both forms not uncommon in thickets.—All islands.

773. Sponia micrantha, Decs.

Fl. April–Sept.   In forests, here and there.—All islands.

774. Ficus crassinervia, Desf.

Fl. Jan.   In forests, not uncommon.—St. Croix (Crequis, Wills Bay).

775. F. trigonata, L.

Fl. May–Aug.   In forests.—St. Croix (rare, Crequis); Virgin Islands (not uncommon).

776. F. lævigata, Vahl.

Fl. Jan.–March.   In forests and on rocks, not uncommon.—St. Croix (Crequis, Jacob's Peak).

777. F. lentiginosa, Vahl.

Fl. May.   In forests on high hills, uncommon.—St. Thomas (Signal Hill).

778. F. populnea, W.

Fl. July–Aug.   Figs geminate in the axils, red with dark spots. On rocks and epiphytic on trees, not uncommon.   Long aërial roots.— All islands.

779. F. pedunculata, Ait.

Fl. Jan.–May.   Figs red, generally inhabited by a small hymenopterous insect.   On rocks, walls, and trees, common.   Long aërial roots.— All islands.

780. Artocarpus incisa, L. (v. Breadfruit-tree).

Fl. May–July.   Fruit not edible.   Naturalized in shady valleys.—All islands.

781. Cecropia peltata, L. (v. Trumpet-tree).

Fl. April–June.   In shady forests, not uncommon.—All islands.

782. Maclura tinctoria, Don (v. Fustic).

Fl. June–Oct. Young shoots with deeply serrate leaves. Wood affording an excellent timber, but now very scarce. In forests, here and there.—All islands.

783. Fleurya æstuans, Gaud.

Fl. June–Dec. On rocks in shady forests, here and there.—St. Croix (Spring Garden); St. Thomas (Crown).

784. Urera elata, Gris.

St. Croix (Spring Garden, West, p. 306; his specimen in Hb. Havn.).

785. U. baccifera, Gaud.

St. Thomas (Wedd. in DC. Prodr. xvi, i, 93).

(West's *Urtica elongata*, Vahl, said, p. 306, to occur in St. Croix, and probably intended for an *Urera*, I have not been able to identify, from want of description and specimens.)

786. Pilea microphylla, Liebm.   *a*), *β*) **trianthemoides**, Lindl., and *γ*) **succulenta** (v. Duck-weed).

Fl. all the year round. On rocks and stones in shady situations. *a*) uncommon; *β*) and *γ*) common.—All islands.

787. P. semidentata, Wedd.

Fl. March–July. Gregarious among rocks on high hills, not uncommon.—St. Thomas (St. Peter).

788. P. grandis, Wedd.

Fl. June. In leaf-mould on high hills, gregarious, uncommon.—St. Thomas (Crown, 1500′).

789. P. nummularifolia, Wedd.

St. Thomas (Hornbeck in Hb. Havn.); Vieques (near Campo Asilo).

790. P. inæqualis, Wedd.

Fl. July–Aug. Gregarious on rocks in forests, uncommon.—St. Thomas (Signal Hill, Crown).

791. P. Sanctæ-Crucis, Liebm. (Vid. Selsk. Skrift., v. Række, ii, 301).

St. Croix (Örsted, l. c.).

792. Rousselia lappulacea, Gaud.

St. Thomas (DC. Prodr. xvi, i, 235; Gris. Fl. p. 160).

[Cultivated species: *Ficus Carica*, L. (v. Fig-tree), and *F. elastica*, L.]

ARISTOLOCHIACEÆ.

793. Aristolochia trilobata, L. (v. Tobacco-pipe).

Fl. May–Aug. On fences and in forests on high hills.—St. Croix (West, p. 305); Virgin Islands (not uncommon).

794. A. anguicida, L. (DC. Prodr. xv, i, 464; Bot. Mag. 4361; Descourtilz, Fl. Méd. des Antilles, iii, 202) (v. Crane's Neck).

Fl. Oct.–Dec. A number of dipterous insects are usually found imprisoned in the lower part of the perigonal tube, whence escape is impossible on account of the downward-bent hairs on the inner surface. The hairs dropping off after fertilization, the imprisoned insects are set at liberty again. In thickets, rare.—St. Croix (Recovery Hill).

BEGONIACEÆ.

795. Begonia humilis, Hort. Kew. (ed. i, vol. iii, 353).

St. Thomas (Finlay in Hb. Mus. Paris. sec. DC. Prodr. xv, i, 297).
[Cultivated occur several species of Begonia.]

AMENTACEÆ.

[Cultivated in gardens and near dwellings: *Casuarina equisetifolia*, Forst. (Fl. June–Aug.) Of very quick growth.]

PIPERACEÆ.

796. Piper Sieberi, Cas. DC. (Enckea, Miq.).

Fl. all the year round. In forests; often gregarious and forming a dense underwood, common. Used for walking-sticks.—All islands.

797. P. Bredemeyeri, Jacq. (Artanthe, Miq.).

Fl. Sept. In shady valleys, not uncommon.—St. Croix (Caledonia, Crequis).

798. P. auritum, Kth.

St. Thomas (DC. Prod. l. c. 321).

799. P. Blattarum, Sprgl.

Fl. Jan.–March. In forests, rare.—St. Thomas (Crown, Signal Hill).

800. P. peltatum, L. (v. Monkey's Hand) (*Potomorphe*, Miq.).

Fl. Feb.–Aug. In forests, along rivulets, and among rocks on high hills.—St. Croix (rare, Caledonia, Springfield); Virgin islands (not uncommon on high hills).

801. Peperomia pellucida, Kth.

Fl. May–Aug. In forests, rare.—St. Croix (Rohrs Minde); St. Thomas (DC. Prod. l. c. 402).

802. P. acuminata, Miq. (*P. guadeloupensis*, Cas. DC.) (v. Stone Ginger).

Fl. all the year round. On rocks in forests, common.—All islands.

803. P. glabella, Dietr.

Fl. May–Sept. In the same places as the preceding, common.—All islands.

804. P. cubana, Cas. DC.

St. Croix (DC. Prod. l. c. 413).

805. P. obtusifolia, Cas. DC., Dietr., Miq. *a*) and *β*) **clusiæfolia.**

Fl. April–July. On rocks and under shady trees in leaf-mould. Gregarious, not uncommon. *a*) all islands ; *β*) St. Thomas (Crown).

806. P. scandens, Ruiz et Pav.

St. Thomas (DC. Prod. l. c. 434).

807. P. polystachya, Miq.

Fl. Dec.–Jan. Stem and lower surface of the leaves reddish. Among rocks in forests, not uncommon, gregarious.—All islands.

B. GYMNOSPERMÆ.

CYCADACEÆ.

[Cultivated in gardens occurs *Cycas revoluta,* Thunb. (v. Sago Palm).]

CONIFERÆ.

[Cultivated in gardens occur several species of Thuja.]

C. MONOCOTYLEDONES.

ALISMACEÆ.

808. Echinodorus cordifolius, Gris.

Fl. April–Aug. Flower expanded only till 10 A. M. Leaves heteromorphous, the primordial ones submerged, linear-lanceolate, passing by degrees into the ordinary emersed ones. In rivulets, here and there.—St. Croix (King's Hill Gut, Armas Hope Gut).

HYDROCHARIDACEÆ, L. C. Rich.

809. Thalassia testudinum, Solander (Koenig).

Not seen flowering. Gregarious in shallow sea-water, very common.—All islands.

POTAMEÆ, Juss.

810. Cymodocea manatorum, Ascherson (Naturf. Freunde in Berlin, Jun.–Oct., 1868).

Not seen flowering, gregarious on the bottom of the sea ; mostly in shallow water.—All islands.

811. Halodule Wrightii, Aschers. (l. c., and Neumayers Anleit. zur wiss. Beob. auf Reisen).

St. Thomas (Krebs sec. Aschers.).

812. Halophila Baillonii, Aschers. (in Neumayer, l. c. p. 367).

Rhizome creeping, thin. Leaves oval, denticulate, whorled or opposite, 3''' long, 1½''' broad. Monœcious.

Fl. ♂ : 3 membranaceous white bracts; 1–3 stamens; filament ⅔''' long; anther cylindrical, yellowish, glabrous, 1-celled. Pollengrains fusiform.

Fl. ♀ : 3 persistent bracts, as in ♂. Ovary sessile, ovate, ⅓''' long, ·loculate. Style bifid, 2¼''' long; branches pointed, often of unequal length. Capsule oval, glabrous, 2''' long; seeds about 20, globose, hard, tessellate on the surface. Starch-grains triangular.

Male flowers very rare compared to the number of female ones.

Fl. all the year round. Gregarious on the bottom of the seaou coarse coral sand in a depth of from two to four fathoms, here and there.—St. Thomas (harbour).

813. Ruppia rostellata, Koch.

Fl. all the year round. Gregarious in shallow rivulets, not uncommon.—St. Croix (King's Hill Gut, in company with a species of Chara); St. Thomas (Tutu Gut, Krebs in Hb. Havn.).

[Another Potamea, possessing a creeping rhizome and delicate linear leaves, has been found by me in the harbour of St. Thomas at a depth of from 3 to 6 fathoms, but on account of only sterile specimens having been obtained it remains as yet undetermined.]

AROIDEÆ.

814. Anthurium Huegelii, Schott (v. Boyer) (*A. acaule,* Sch.).

Fl. July–March. Young radical leaves very different from the older ones, being lanceolate and long-petioled. I consider Schott's *A. acaule* not to be specifically distinct from this species. On rocks and trees, not uncommon, often gregarious.—All islands.

815. A. macrophyllum, Sch.

Fl. July-Dec. Among rocks in forests, uncommon. St. Jan (near Bethania).

816. A. cordifolium, Kth. (v. Wild Tanier, Maroon Jancole) (Bot. Mag. t. 2801, 5801 being a misprint in Gris. Fl. p. 508).

Fl. July–Nov. Among rocks in forests, gregarious, here and there.—St. Croix (Wills Bay, Blue Mountain); St. Jan (Macumbi, 1200').

817. Dieffenbachia Seguine, Sch. (v. Dumb Cane).

Fl. May–Dec. In moist places on high hills, uncommon.—St. Thomas (Caret Bay, 1000').

818. Philodendron hederaceum, Sch.

Fl. Aug. On trees in dense forests, rare.—St. Thomas (Crown, 1400′).

819. Ph. giganteum, Sch. (Prod. Syst. Aroid. p. 261).

Fl. March–July. Petiole 2′–2½′ long; lamina 2½′–3′ long, 2′ broad. Peduncle 1½″–3½″ long; spathe 11″–12″ long, opening itself only during two nights. Spadix white, giving out a strong odour and considerable high temperature during anthesis. Numerous aërial roots, stem 1′–2′ long. Among rocks in dense forests on high hills, gregarious on trees.— St. Thomas (Signal Hill and Crown, 1500′).

(The picture in Bot. Mag. t. 3314, of the much smaller *Ph. fragrantissimum*, Kth. (*Caladium*, Hook.), gives a good representation of the habit of this species.)

820. Caladium smaragdinum, C. Koch (Schott, l. c. 165) (v. Guinea Ginger).

Fl. May–July. Rhizome tuberous, yellow. In pastures on high hills, not uncommon.—St. Thomas (Signal Hill, above St. Peter, 1400′).

821. Xanthosoma atrovirens, C. Koch (v. Scratch-throat).

Not seen flowering. Rhizome large, tuberous, used as a vegetable. Leaves pungent when eaten as spinach. Cultivated and naturalized on provision grounds.—St. Croix; St. Thomas.

822. X. sagittæfolium, Sch. (v. Tanier).

Fl. July. Lamina of the spathe white, with a delicate rosy tinge. Spathe disclosing itself during two nights from 7 to 10 o'clock; spadix meanwhile giving forth a strong fragrance and showing a temperature of 12° C. above that of the air. Leaves used as spinach and the tuberous rhizome as a common vegetable. Cultivated and naturalized on provision grounds.—All islands.

823. X. ? hastatum, Egg. (Arum, Vahl.) (v. Indian Kale).

Not seen flowering. Leaves hastate, with long pointed oblique basilar lobes; used for spinach. (Naturalized?) Cultivated and spontaneous in forests.—All islands.

824. Pistia occidentalis, Bl.

Fl. all the year round. Cultivated and naturalized in gardens.—St. Thomas.

825. Lemna minor, L.

Not seen flowering. In rivulets, not uncommon.—St. Croix (Jealousy Gut, Fair Plain Gut).

[Cultivated species: *Caladium bicolor*, Vent., *C. pictum*, DC., and *C. picturatum*, Linden.]

TYPHACEÆ.

826. Typha angustifolia, L., var. **domingensis**, Pers.

Fl. Sept.–March. Used for making mats. In rivulets and around lagoons, not uncommon.—St. Croix; St. Jan.

PANDANACEÆ.

[Cultivated in gardens occurs *Pandanus odoratissimus*, L. fil. (v. Screw Pine).]

PALMÆ.

827. Thrinax argentea, Lodd. (v. Teyer-tree).

Fl. May–June. Stem 10′–20′ high. Leaves used for making ropes, thatching roofs, and other domestic purposes. On the northern slope of the hills in forests and tickets.—St. Croix (very rare, only one specimen seen, near Bellevue Mill); Virgin Islands (common).

828. Oreodoxa regia, Kth. (v. Mountain Cabbage).

Fl. April–Aug. The young leaf-bud used as cabbage. Berries eaten by hogs. In forests and along roads, common.—All islands.

829. Cocos nucifera, L. (v. Cocoa-nut Tree).

Fl. Feb.–March. Leaves used for thatching roofs. The ripe fruit, although occurring in abundance, is scarcely used, and of no economical importance. Naturalized along the seashore and along roads.—All islands.

COMMELYNACEÆ.

830. Tradescantia geniculata, Jacq. β) **effusa**, Mart.

Fl. March. Seeds bluish, verruculose.—Vieques (near Campo Asilo).

831. T. zebrina, Hortul. (v. Wandering Jew).

Fl. May. Naturalized in gardens.—St. Croix; St. Thomas.

832. T. discolor, Sw.

Fl. April–Aug. Stamens often by retrograde metamorphosis transformed into petals. Naturalized in waste places and near dwellings.—All islands.

833. Callisia repens, L.

Fl. Jan.–March. Gregarious in shady places, not uncommon.—All islands.

834. C. umbellulata, Lam.

Fl. Jan. Seeds white with a red spot. Flowers monandrous. Among rocks in shady places, rare.—St. Thomas (Signal Hill, 1200′).

835. Commelyna cayennensis, Rich. (French Grass).

Fl. all the year round. Flower expanded till 9 A. M. One of the sterile stamens always abortive. In moist localities, common.—All islands.

836. C. elegans, Kth. (v. French Grass).

Fl. all the year round. Flower ephemeral. In moist localities, very common.—All islands.

GRAMINACEÆ.

837. Bambusa vulgaris, Schrad. (v. Bamboo Cane).

Not seen flowering. Naturalized along rivulets and in gardens.— St. Croix; St. Thomas.

838. Arthrostylidium capillifolium, Gris. (Plant. Wright. in Mem. Amer. Acad. viii, 531, 1862).

Not seen flowering. In forests, climbing among trees and shrubs to a considerable height, rare.—St. Thomas (Flag Hill, 700′); St. Jan (Hornbeck in Hb. Havn., from "a large cataract, called Battery"); Vieques (flowering specimens from Hornbeck in Hb. Havn.; others received from Campo Asilo by me).

839. Eragrostis poæoides, P. Br.

Fl. June–Dec. Stigmas white. Along roads and in dry localities, often gregarious, common.—St. Croix; St. Thomas.

840. E. ciliaris, Lk.

Fl. March–Dec. Anthers black. In dry localities, common.—All islands.

841. Sporobolus virginicus, Kth. (v. Shander).

Fl. May–Oct. Anthers and stigmas yellow. Used in baths for children. Along the coast and lagoons, common.—All islands.

842. S. litoralis, Kth. (v. Shander).

Fl. May–Dec. In the same places as the preceding, common.—All islands.

843. S. indicus, R. Br. (v. Hair-grass).

Fl. May–Oct. Anthers purple; stigmas yellow. Along roads and ditches.—All islands.

844. Aristida stricta, Mich.

Fl. March–Dec. Anthers yellow. Awns of unequal length, always longer than the glumes. Along ditches and in thickets, here and there.—St. Croix (Crequis, Fair Plain); St. Thomas (Schl.); St. Jan (Adrian Estate).

845. Olyra latifolia, L. β) **arundinacea**.

Fl. Dec.–Jan. In forests, rare.—St. Jan (Cinnamon Bay); Vieques (Campo Asilo).

846. Pharus glaber, Kth.

Fl. June–Dec. Anthers yellow; stigmas white. In forests, not uncommon.—All islands.

847. Pappophorum alopecuroides, Vahl.

Fl. Feb.–March.   1'–3' high.   Among rocks near the coast, rare.—
Buck Island, near St. Thomas; Virgin Gorda (Vahl in Symb. Bot. iii, 10).

848. Bouteloua litigiosa, Lag.

Fl. Oct.–Jan.   Anthers red; stigmas white.   In thickets and waste
places, not uncommon.—St. Thomas (Cowell's Hill—Town).

849. Leptochloa mucronata, Kth.

Fl. May–Oct.   Spikelets often 1-flowered.   Along ditches, not uncommon.—St. Croix.

850. L. virgata, P. Br.   *a*), *β*) **gracilis**, Ns., and *γ*) **multiflora**, Egg.

Fl. May–Dec.   Anthers white; stigmas purple.   *γ*) spikelets 9-flow-
ered.   Awns very short; fertile glumes not ciliate.   Along roads, common.—*a*) and *β*) all islands; *γ*) St. Croix (Work and Rest).

851. Chloris eleusinoides, Gris.

Fl. May–Nov.   Along ditches, here and there.—St. Croix (Beeston
Hill, Mount Welcome).

852. Ch. radiata, Sw.

Fl. May–Oct.   Stigmas brown.   Gregarious along roads, common.—
All islands.

853. Ch. ciliata, Sw.

Fl. Feb.–Sept.   Anthers rosy.   My specimens show only one sterile
flower in each spikelet besides the fertile one (see Swartz's Flora Ind.
Occ. p. 189).   Along roads, not uncommon.—All islands.

854. Dactyloctenium ægyptiacum, W. (v. Ten-per-cent Grass).

Fl. March–Nov.   Anthers straw-coloured; stigmas white.   A good
pasture-grass.   Along roads and in fields, common.—All islands.

855. Eleusine indica, L.

Fl. March–Dec.   Anthers greyish; stigmas purple.   Common everywhere.—All islands.

856. Cynodon Dactylon, Pers. (v. Bay Grass, Billy Grass).

Fl. May–Oct.   Anthers straw-coloured, with purple spots; stigmas
purple.   A good pasture-grass, and fit for making good hay, but at the
same time a most troublesome weed in cane-fields on account of its long
and creeping rhizome.   Said to have been introduced.   Along the coast
and in fields, gregarious.—St. Croix and St. Thomas (very common); St.
Jan (uncommon, Little Plantation).

857. Paspalum compressum, Ns. (v. Flat Grass).

Fl. June–Oct. Anthers light yellow; stigmas white. Near ditches and in shady localities, not uncommon.—All islands.

858. P. conjugatum, Berg.

Fl. June–Dec. Anthers yellow; stigmas white. In moist localities, common.—All islands.

859. P. pusillum, Vent.

St. Thomas (Flügge sec. Gris. Syst. Unt., p. 114).

860. P. distichum, L. *a*) and *β*) **vaginatum,** Sw.

Fl. June–Aug. Proterandrous. Anthers light yellow; stigmas black. Along rivulets, not uncommon.—St. Croix; St. Thomas.

861. P. notatum, Flügge.

St. Thomas (Flügge sec. Gris. Syst. Unt., p. 114).

862. P. cæspitosum, Flügge.

Fl. May–Sept. Anthers orange-coloured. In moist localities, not uncommon.—All islands.

863. P. glabrum, Poir.

Fl. May–July. Here and there along ditches.—St. Thomas (Schl.); St. Jan (Riff Bay).

864. P. plicatulum, Michx.

Fl. March–Sept. Along the seacoast, not uncommon.—All islands.

865. P. virgatum, L. *a*).

Fl. May–Oct. Anthers straw-coloured; stigmas white. In moist localities, not uncommon.—All islands.

866. P. paniculatum, L.

St. Thomas (Schlechtendal).

867. P. spathaceum, HB. K.

St. Thomas (Schlechtendal).

868. Digitaria filiformis, Mühl.

Fl. Dec. In dry thickets, here and there.—St. Thomas (Cowell's Hill).

869. D. marginata, Lk. (v. Running Grass).

Fl. March–Sept. Anthers purple with white stripes; stigmas purple. A good pasture-grass. Along ditches and roads, common.—All islands.

870. D. setigera, Kunth.

Fl. June–Oct. Anthers and stigmas purple. Along roads, common.—
All islands.

871. Eriochloa punctata, Hamilt.

Fl. March–Sept. Anthers brownish; stigmas black. In moist localities, here and there.—St. Croix (Crequis, La Grange); St. Thomas
(Schl.).

872. Stenotaphrum americanum, Schrank (v. Horse Grass).

Fl. May–Aug. Anthers orange-coloured; stigmas purple. Along the
coast and in moist localities, gregarious, common.—All islands.

873. Orthopogon setarius, Spreng.

Fl. March–Dec. Anthers light purple; stigmas purple. In forests,
common.—All islands.

874. Panicum paspaloides, Pers.

Fl. March–Sept. Anthers reddish; stigmas straw-coloured. The
hermaphrodite flower in this and all other species of Panicum is proterandrous, the stamens dropping off before the stigmas appear. These
latter are then fertilized by the agency of the wind from other individuals before the stamens of the male flower make their appearance, self-
fertilization being thus evidently impossible. Along rivulets and in
moist localities, not uncommon.—St. Croix; St. Thomas.

875. P. brizoides, L.

St. Thomas (Schlechtendal).

876. P. colonum, L.

Fl. March–Sept. Anthers purple; stigmas black. Along roads and
ditches, common.—All islands.

877. P. prostratum, Lam.   α) and β) **pilosa,** Egg.

Fl. June–July. Anthers orange-coloured; stigmas black.   β) Rhachis
of spikelets pilose.—α) All islands (common); β) St. Croix (La Grange).

878. P. fuscum, Sw. (v. Sour Grass).   α) and β) **fasciculatum,** Sw.

Fl. Feb.–Sept. Anthers orange-coloured; stigmas purple. Abhorred
by the cattle.—α) All islands.   β) St. Croix; St. Thomas (Schlechtendal). Not uncommon.

879. P. molle, Sw. (v. Yerba de Pará, Spanish Grass).

Fl. May–Oct. Anthers yellow; stigmas purple. Naturalized here
and there in pastures.—St. Croix (Cotton Grove).

880. P. diffusum, Sw.

Fl. May–Oct. Anthers orange-coloured; stigmas dark purple. In moist localities, uncommon.—All islands.

881. P. maximum, Jacq. (v. Guinea Grass) (*P. polygamum*, Sw.).

Fl. June–Sept. Anthers brownish; stigmas light purple. A splendid pasture-grass, growing to the height of 12′, forming dense tufts and being propagated by the rhizome. Naturalized and cultivated everywhere.—All islands.

882. P. divaricatum, L. α) and β) puberulum.

Fl. May–Dec. Anthers light yellow; stigmas white. Resembling a thin Bamboo Cane. 8′–16′ high. Both forms not uncommon in forests, climbing over trees and shrubs.—All islands.

883. P. glutinosum, Sw.

St. Croix (West, p. 267).

884. P. brevifolium, L.

Fl. Aug.–Dec. Anthers and stigmas white. In gardens and along roads, here and there.—St. Thomas (Barracks).

885. P. cayennense, Lam.

St. Thomas (Schlechtendal).

886. Setaria glauca, P. Br. α).

Fl. May–Oct. In forests, common.—All islands.

887. S. setosa, P. Br. a) and β) caudata, R. S. (v. Sour Grass).

Fl. April–Dec. Anthers orange-coloured; stigmas purple. a) 3′–7′ high; in forests and along ditches, common.—All islands. β) in dry thickets, uncommon.—St. Thomas (Cowell's Hill).

888. Cenchrus echinatus, L. β) viridis, Spreng. (v. Burr Grass).

Fl. April–Dec. Anthers light yellow; stigmas white, with a purple spot in the middle. The ripe farinaceous seeds eaten by the cattle. Along the coast, very common.—All islands.

889. Anthephora elegans, Schreb.

Fl. Jan.–Oct. Anthers brownish. In thickets, here and there.—St. Croix; St. Thomas.

890. Tricholæna insularis, Gris. (v. Bitter Grass, Long Grass).

Fl. March–Dec. Anthers brownish; stigmas white. Never touched by cattle whilst green, on account of its bitter taste. Spikelets easily detached and carried far away by the wind. Very common along roads and in dry places.—All islands.

891. Lappago aliena, Spreng.

Fl. May–Dec. Stigmas white. Generally both spikelets fertile. Near ditches and in thickets, common.—All islands.

892. Andropogon saccharoides, L.

Fl. Aug.–Oct. Anthers light yellow; stigmas dark purple. Awn not twisted. Along roads, here and there.—St. Croix (Beeston Hill Grange).

893. Anatherum bicorne, P. Br. (v. Jolly Grass).

Fl. July–Oct. 2'–4' high. Used for thatching roofs. Not eaten by the cattle. Gregarious on high hills, where it is difficult to counteract its spreading, even by burning it now and then.—St. Thomas (northern slope of the highest ridge).

894. Sorghum vulgare, Pers. (v. Guinea Corn).

Fl. Dec. 8'–16' high. Naturalized and cultivated for herbage and for making flour of the grain.—All islands, principally St. Croix and Vieques.

895. Saccharum officinarum, L. (v. Sugar-cane).

Fl. Dec.–May. Naturalized and cultivated. Sugar-growing islands are now only two, viz., St. Croix and Vieques, whilst the other Virgin Islands have only a very few cane estates, principally for selling the raw cane in the markets. The average produce of sugar from both the above-mentioned islands is about 25 million pounds. The plant is propagated by cuttings that are laid entirely under ground.

(The genus *Panicum* excepted, all *Graminaceæ* are proterogynous.)

[Cultivated species: *Andropogon Schœnanthus*, L. (v. Lemon-grass), *Zea Mays*, L. (v. Indian Corn), and *Coix Lacryma*, L. (v. Job's Tears).]

CYPERACEÆ.

896. Cyperus polystachyus, Rottb.

Fl. July. On high hills, rare.—St. Thomas (Crown, 1500').

897. C. lævigatus, L. (Cod. p. 61) (*C. mucronatus*, Rottb.). *a)* **albidus.**

Fl. March–Oct. Connective pointed. Along rivulets, not uncommon.—St. Croix; St. Thomas (Schl., Böckeler).

898. C. compressus, L.

Fl. Dec. Flowers 2-androus. Near the coast in moist places, uncommon.—St. Thomas (Haven Sight).

899. C. confertus, Sw.

Fl. Dec. In thickets, here and there.—St. Thomas (Cowell's Hill); St. Croix (Gris. Fl. 563).

900. C. ochraceus, Vahl.

Fl. May–Oct. In moist localities, uncommon.—St. Croix (Crequis).

901. C. viscosus, Ait.

Fl. April–Nov. Stamens always 3 (see Swartz's Fl. Ind. Occ. p. 113). Seeds germinating in moist weather on the parent, and often growing out into young plants an inch or two in length. Along rivulets and ditches, not uncommon.—St. Croix; St. Thomas.

902. C. surinamensis, Rottb.

St. Thomas (Schl.).

903. C. articulatus, L. (v. Sting Bisom).

Fl. March–Sept. In ditches, not uncommon.—St. Croix; St. Thomas.

904. C. rotundus, L. (v. Nut Grass).

Fl. all the year round. Tubers sweet, eaten by hogs. A troublesome weed, very common in fields and along roads.—All islands.

905. C. brunneus, Sw. (*C. planifolius*, Rich.).

Fl. May. On the coast and near lagoons, not uncommon.—All islands.

906. C. sphacelatus, Rottb.

Fl. Feb. On high hills in pastures, uncommon.—St. Thomas (Signal Hill).

907. C. distans, L.

Fl. Aug. In pastures on high hills, common.—St. Thomas (Signal Hill).

908. C. unifolius, Boeckler (Linnæa, Neue Folge, ii, 374).

St. Croix (Ravn in Reliq. Lehm.).

909. C. filiformis, Sw.

Fl. all the year round. In moist localities, not uncommon.—St. Thomas.

910. C. odoratus, L.

Fl. April–Oct. Near rivulets and ditches, here and there.—St. Croix (Mount Pleasant, Annas Hope).

911. C. pennatus, Lam. (Boeckler, l. c. 404) (*C. Ehrenbergii*, Kth., *C. flexuosus*, Vahl).

Fl. all the year round. Along the coast, not uncommon.—St. Thomas.

912. C. ligularis, L.

Fl. March–Dec. Along rivulets, not uncommon.—All islands.

913. C. flavomariscus, Gris. (*C. flavus*, Bœckler).

Fl. Aug. In pastures on hills, here and there.—St. Thomas (Signal Hill); Buck Island (near St. Thomas).

914. Kyllinga filiformis, Sw. *a*) and *γ*) **capillaris**, Gris.

Fl. June–Dec. Involucral leaves of various lengths. Both forms not uncommon in forests.—St. Croix (The William, Eliza's Retreat).

915. K. triceps, Rottb.

Fl. March. In shady moist localities.—St. Jan (Baas Gut).

916. K. monocephala, Rottb.

Fl. all the year round. In moist places in forests, common.—All islands.

917. K. brevifolia, Rottb. (Emend. in Bœckler, Linnæa, 1867, 425).   *β*) **longifolia**.

St. Thomas (Ehrenberg sec. Bœckler).

918. Scirpus capitatus, L.

Fl. all the year round. Achenium black. Along rivulets, common.—All islands.

919. S. nodulosus, Kth.

Fl. March–Dec. Along rivulets and in ditches, uncommon.—St. Croix (Adventure).

920. S. subdistichus, Bœckler (Linnæa, 1869–70, 490).

St. Thomas (Beklr.).

921. S. mutatus, Vahl.

Fl. March–Dec. Filaments flat; style often bifid. In moist places, not uncommon.—St. Croix; St. Jan.

922. S. ferrugineus, L.

Fl. all the year round. Filaments flat. Gregarious in tufts on the sandy seashore and near lagoons, uncommon.—St. Croix (Frederiksted); St. Jan (Reef Bay).

923. S. brizoides, Sw. (*Fimbristylis polymorpha*, Bœckler).

Fl. Aug.–Sept. In pastures on high hills, common.—Virgin Islands.

924. Rhynchospora pusilla, Gris.

Fl. Feb.–July. Anthers 1¼′′′ long. In pastures on hills, rare.—St. Thomas (Signal Hill, 1400′).

925. R. pura, Gris.

Fl. Feb.–Aug. Seeds often germinating on the parent. In the same places as the preceding. St. Thomas (Signal Hill).

926. Scleria pratensis, Lindl. (v. Cutting Grass).

Fl. April–Nov. In forests and pastures on high hills, uncommon.—St. Croix (Springfield, Mount Eagle); St. Thomas (Signal Hill).

927. S. scindens, Ns. (v. Razor-grass).

Fl. Aug.–Sept. In forests, rare.—St. Thomas (Signal Hill, 1500′).

928. S. filiformis, Sw. (*S. lithosperma*, W.).

Fl. May–Nov. In thickets, not uncommon.—St. Croix (King's Hill); St. Thomas (Cowell's Hill).

[All *Cyperaceæ* are proterogynous, with white stigmas and light yellow anthers.]

LILIACEÆ.

929. Aloe vulgaris, L. (v. Sempervivie).

Fl. March–April. Gregarious on limestone (naturalized?), common.—All islands.

930. Yucca gloriosa, L.

Fl. June–Aug. Naturalized in gardens and near dwellings.—St. Croix; St. Thomas.

931. Agave americana, L. (v. Karatá).

Fl. Feb.–May. On dry hills, common.—All islands.

932. A. sobolifera, Salm-Dyck. (v. Karatá).

Very seldom or never bearing flowers. Propagated by bulblets in June–July, growing out to a considerable size whilst still on the parent. On hills and in thickets, not uncommon.—All islands.

933. Fourcroya cubensis, Haw. (v. Female Karatá).

Fl. March and July–Aug. In dry thickets, not uncommon.—St. Croix; St. Thomas.

934. Pancratium caribæum, L. (v. White Lily, Ladybus).

Fl. May–Nov. Flowers nocturnal; fragrant. On rocky coasts, not uncommon.—All islands.

935. Crinum erubescens, Ait.

Fl. all the year round. Flowers nocturnal; fragrant. Along rivulets, here and there.—St. Croix (Høgensborg).

936. Amaryllis equestris, Ait. (v. Red Lily).

Fl. March–Oct. On rocky shores, gregarious, not uncommon.—All islands.

937. A. tubispatha, Ker. (v. Snow-drop).

Fl. April–Oct., especially after heavy rains. In fields and near dwellings, not uncommon.—All islands.

[Cultivated species: *Allium fistulosum*, L. (v. Ciboule), *Polyanthes tuberosa*, L. (v. Tuberose), and *Crinum giganteum*, Andr.]

ASPARAGINACEÆ.

938. Sanseviera guineensis, W. (Spec. ii, 159) (Bot. Mag. t. 1179) (v. Guana-tail).

Fl. Nov.–Dec. Fibres of the leaves yield a good material for ropes. Naturalized here and there on dry hills, gregarious.—St. Croix (Friedensfeld); St. Thomas (around town).

SMILACEÆ.

939. Smilax havanensis, Jacq.

Not seen flowering. In forests, here and there.—St. Croix (Caledonia, Wills Bay, Rohr's Minde).

940. S. populnea, Kth. (Enum. Plant. v, 192).

Fl. June–July (♂). Unarmed. Leaves 4″–5″ long, 3″–4″ broad. In forests, a high climber, rare.—St. Thomas (Flag Hill, 960′).

DIOSCOREACEÆ.

941. Dioscorea pilosiuscula, Bert.

Fl. Dec., but rarely. Older leaves purple beneath, broad white stripes on the upper surface. Male inflorescence 3″ long, pendulous. Axillar bulbs large, often bifid, greyish-brown, generally producing leaves whilst still in connection with the parent, dropping off later and forming new plants. In shady forests, uncommon.—St. Thomas (Signal Hill, northern slope above St. Peter, 1000′).

942. D. alata, L. (v. Yam). a), β) **vulgaris**, Miq.

Not seen flowering. Propagated by the rhizome. Naturalized and cultivated in provision grounds. Rhizome affording a nutritive vegetable.—All islands.

943. D. altissima, Lam. (v. Yam).

Not seen flowering. Stem cylindrical. Occurring in the same places and used in the same way as the preceding.—All islands.

944. Rajania pleioneura, Gris.

Fl. Dec. In forests, rare.—St. Thomas (Flag Hill, 800′).

945. R. hastata, L.

Fl. Sept.–Dec. In forests and on fences on high hills, not uncommon.—St. Thomas (Signal Hill, Northside) (St. Croix?).

IRIDACEÆ.

946. Cipura plicata, Gris. (v. St. Jan Grass, Bloodroot).

Fl. all the year round. Bulbs crimson. Naturalized in gardens and valleys.—All islands.

BROMELIACEÆ.

947. Bromelia Pinguin, L. (v. Pinguin).

Fl. Dec. and April–June. Pulp edible, acid. Used for fences. Gregarious in forests and thickets, common.—All islands.

948. Chevalliera lingulata, Gris.

Fl. March–July. Petals white, with a bluish point. Berry glabrous, pink or blue. On trees and rocks on high hills, not uncommon.—St. Thomas (Crown, Signal Hill, 1500′); St. Jan (Macumbi).

949. Pitcairnia angustifolia, Ait.

Fl. Aug.–Sept. Seeds red, pointed at the base; appendage white, truncate. On trees and rocks.—St. Croix (rare, King's Hill Gut); Virgin Islands (common, especially on the coast).

950. Tillandsia fasciculata, Sw.

Fl. Jan.–Feb. Capsule a little shorter than the bract. On trees in forests and on high hills, uncommon.—St. Thomas (Crown); St. Jan (Baas Gut).

951. T. utriculata, L. (v. Wild Pine).

Fl. Feb.–Aug. Inflorescence over 8′ high. On trees and rocks, common.—All islands.

952. T. recurvata, L. (v. Old Man's Beard).

Fl. Jan.–Feb., but very rarely. Seeds often germinating in the capsule. Used for stuffing mattresses. On trees, gregarious, very common.—All islands.

953. T. usnecides, L. (v. Old Man's Beard).

Fl. March, rarely. Petals greenish. On shrubs, common, gregarious.—All islands.

954. Catopsis nutans, Gris.

Fl. June–Aug. Petals fleshy, white. Seeds brown; pappus 1¼″

long, white, silky. On trees and rocks on high hills, not uncommon.—
St. Thomas (Signal Hill, Crown, 1400′–1500′).

[Cultivated species: *Ananassa sativa*, Lindl. (v. Pine-apple).

MUSACEÆ.

955. Musa paradisiaca, L. (v. Plantain).

Fl. May–Aug. Fruit eaten only boiled or fried. Naturalized and
cultivated, but rare.—All islands.

956. M. sapientium, L. (v. Banana).

Fl. May–Nov. Fruit eaten raw or fried. Naturalized and cultivated
everywhere, occurring in several varieties (Bacuba, Fig, Lady-finger,
St. Vincent Banana, etc.).—All islands.

SCITAMINEÆ.

957. Renealmia sylvestris, Gris.

Fl. Aug. In forests in shady and moist localities, rare.—St. Croix
(Golden Rock); St. Thomas (Signal Hill, 1400′).

958. Zingiber officinalis, Rosc. (v. Ginger).

Fl. Sept. Naturalized and cultivated in forest districts, here and
there.—St. Croix; St. Thomas.

959. Canna indica, L. (v. Indian Shot).

Fl. all the year round. In moist places and near dwellings, not uncommon.—All islands.

960. C. Lamberti, Lindl. (v. Scarlet Indian Shot).

Fl. all the year round. Naturalized in gardens.—All islands.

961. C. edulis, Ker. (v. Tout-le-mois).

Fl. all the year round. Tubers used for producing salep. Naturalized and cultivated along rivulets.—All islands.

962. Maranta arundinacea, L. (v. Arrow-root).

Not seen flowering. Tubers yielding the best kind of salep. Naturalized and cultivated here and there.—All islands.

[Cultivated species: *Alpinia nutans*, Raf. (v. Shell-plant), and *Cur-
cuma longa*, L. (v. Turmeric).

ORCHIDACEÆ.

963. Liparis elata, Lindl.

Fl. June–Dec. Bracts purple. My specimens on the whole somewhat smaller than the picture in Bot. Mag. t. 1175. On red clay among

rocks on high hills, here and there.—St. Thomas (Liliendal, Bonne Resolution).

964. Epidendrum subæquale, Eggers, n. sp.

Fl. Feb.-March. Tubers cylindrical, small, several-leaved. Leaves 2–5, linear, channelled, pointed, much shorter than the scape; sterile bracts short, distant, pointed, floral ones smaller; flowers in a simple raceme, 3–4. Perigonial divisions lanceolate, pointed, nearly conform. Lip slightly adnate to the column, 3-lobed; lobes rounded, the two lateral ones a little shorter than the middle one. Column auricled below the anther; auricles small, purple. Ovary linear, striate, ½″ long. Allied to *E. aciculare*, Batem., but leaves several, much shorter than the scape, and lip broadly 3-lobed. Leaves 5″–6″ long, 2‴ broad; scape 20″–24″ high, straight. Peduncles ½″ long; perigonial divisions greenish, with brown spots, ½″ long; lip purple, with darker stripes and a yellow crest in the middle, ½″ long. The whole plant of a sometimes darker, sometimes lighter hue, flowers even sometimes quite white. On rocks and the roots of trees in dry thickets, here and there.—St. Thomas (Cowell's Hill, Solberg).

965. E. bifidum, Aubl.

Fl. May–Dec. On trees and rocks, not uncommon.—All islands.

966. E. ciliare, L.

Fl. June–Feb. Flowers fragrant. Gregarious on rocks and old treetrunks, common.—All islands.

967. E. cochleatum, L. (Bot. Mag. t. 151, bad).

Fl. April–May. On trees in forests, rare.—St. Croix (Mount Eagle, 1150′; Jacob's Peak, 950′).

968. E. patens, Sw.

Fl. July–Aug. Leaves distichous; scape compressed, 1′–2′ high. On rocks in leaf-mould, rare, on high hills.—St. Thomas (Signal Hill, 1500′).

969. Brassavola cucullata, R. Br.

Fl. June–Octb. Gregarious on rocks, rare.—St. Thomas (John Bruce Bay).

970. Polystachya luteola, Hook.

Fl. March–Nov. Flowers often cleistogamous and normal on the same branch and at the same time. Both forms yielding good seeds. On rocks and old tree-trunks, not uncommon on hills.—St. Thomas (Signal Hill, 1200′–1500′).

971. Oncidium Lemonianum, Lindl.

Fl. May–July. Never giving fruit, but propagating itself by producing young plants from buds in the axils of the sterile bracts below the flowers, which remain in connection with the parent plant, and thus often forming long colonies of plants from one tree to another. In forests and thickets, gregarious, but rare.—St. Thomas (Picara Peninsula, Fortuna).

(The lateral sepals in my specimens being distinct, I am inclined to retain Lindley's specific name instead of uniting my plant with *O. tetrapetalum*, W., as done by Grisebach.)

972. O. variegatum, Sw.

Fl. July–Octb. On rocks and trees in shady places, not uncommon.—Virgin Islands.

973. Prescottia myosurus, G. Rchb.

Fl. March. In grass-fields on high hills, uncommon.—St. Thomas (Signal Hill, 1400′).

974. Spiranthes elata, Rich.

Fl. March. Leaves deciduous during anthesis. In leaf-mould on high hills, not uncommon.—Virgin Islands.

975. Stenorrhynchus lanceolatus, Rich.

Fl. May. Leaves deciduous during anthesis. Only $\frac{1}{2}$′–1′ high. In clayey soil among rocks on high hills, rare.—St. Thomas (Signal Hill, Crown).

976. Habenaria maculosa, Lindl.

Fl. Feb. Spur 1″ long, nectariferous. In pastures on high hills, rare.—St. Thomas (Signal Hill).

977. H. alata, Hook.

Fl. Feb. Spur 6‴ long, nectariferous. In the same localities as the preceding, rare.—St. Thomas (Signal Hill, above St. Peter, 1400′).

II. CRYPTOGAMÆ VASCULARES.

LYCOPODIACEÆ.

978. Lycopodium cernuum, L.

Gregarious among rocks on high hills, here and there.—St. Thomas (Crown, Signal Hill).

979. Psilotum triquetrum, Sw.

In shady places among rocks, not uncommon.—St. Croix (Crequis); St. Thomas (Signal Hill).

FILICES.

980. Ophioglossum reticulatum, L.

In pastures under rocks on high hills, not uncommon.—St. Thomas (Crown).

981. Davallia aculeata, Sw. (v. Prickly Fern).

In pastures on high hills, here and there.—St. Thomas (Signal Hill, above St. Peter, 1300′).

982. Adiantum villosum, L.

Among rocks in forests, uncommon.—St. Croix (Crequis, Vieques).

983. A. intermedium, Sw.

On high hills, not uncommon.—St. Thomas (Signal Hill).

984. A. microphyllum, Kaulf.

Fragrant in the morning. In dense forests, uncommon.—St. Thomas (Crown).

985. A. tenerum, Sw. (v. Maiden-hair).

In thickets, not uncommon.—All islands.

986. A. fragile, Sw.

In the same localities as the preceding, uncommon.—All islands.

987. Cheilanthes microphylla, Sw.

St. Croix (West, p. 313, Benzon in Hb. Havn.); St. Thomas (Ravn in Hb. Havn.).

988. Pteris longifolia, L.

Along rivulets in forests, rare.—St. Croix (Crequis).

989. P. pedata, L.

Gregarious in forests, here and there.—St. Thomas (Signal Hill, near St. Peter).

990. Tænitis lanceolata, R. Br.

In leaf-mould on rocks, not uncommon.—All islands.

991. Antrophyum lineatum, Kaulf.

In forests, rare.—St. Thomas (St. Peter).

992. Blechnum occidentale, L.

Gregarious in pastures and forests, very common.—All islands.

993. Chrysodium vulgare, Fée.

In marshy soil, gregarious; up to 15' high. Not uncommon.—All islands.

994. Hemionitis palmata, L. (v. Strawberry Fern).

Propagating itself by buds from the serratures of the frond. Gregarious in shady forests, here and there.—St. Croix (Eliza's Retreat); St. Jan (Rogiers, King's Hill).

995. Gymnogramme calomelanos, Kaulf. (v. Silvery Fern).

On hills and among stones, not uncommon.—All islands.

Var. pumila, Egg.

Dwarfy, cartilaginous. On old walls, here and there.—St. Croix (Bodkin); St. Thomas (Cowell's Battery).

996. Asplenium serratum, L.

Frond up to 4' long. On rocks in forests, very rare.—St. Thomas (Signal Hill, 1400').

997. A. firmum, Kze.

St. Thomas (Griseb. Syst. Unters. p. 134) (A. abscissum, W.).

998. A. pumilum, Sw.

On clayey soil in forests, gregarious, here and there.—St. Thomas (Matthis Gut); St. Jan (Rogiers).

999. Aspidium punctulatum, Sw.

In forests, not uncommon.—St. Thomas.

1000. A. semicordatum, Sw.

In shady localities, not uncommon.—Virgin Islands.

1001. A. patens, Sw.

In forests, here and there.—St. Croix (Crequis); St. Thomas (Crown).

1002. A. molle, Sw.

In the same localities as the preceding, not uncommon.—St. Thomas (Signal Hill).

1003. A. invisum, Sw. α).

In shady localities, rare.—St. Croix (Crequis).

1004. Polypodium tetragonum, Sw.

In forests, not uncommon.—All islands

1005. P. crenatum, Sw.

St. Croix (West, p. 313, Benzon in Hb. Havn.); St. Thomas (Hb. Havn.).

1006. P. aureum, L.

On dead trees and rocks, not uncommon.—All islands.

1007. P. areolatum, Thunb.

In the same places as the preceding, but rare.—St. Thomas (Crown).

1008. P. incisum, Sw.

St. Croix (West, p. 313).

1009. P. incanum, Sw.

Among roots of large trees, gregarious, not uncommon. All islands.

1010. P. piloselloides, L.

In forests and pastures among rocks on high hills, here and there.—St. Thomas (Signal Hill, 1300′).

1011. P. serpens, Sw.

On trees and rocks on high hills, rare.—St. Croix (top of Mount Eagle, 1150′).

1012. P. Phyllitidis, L. α) and β) repens.

In forests on rocks and trees, not uncommon.—All islands.

1013. Cyathea arborea, Sw.

Stem 12′–15′ high, 3″ diam. In forests on high hills, rare.—St. Thomas (Crown, western slope, 1400′; Caret Bay Gut).

CORRECTIONS AND ADDITIONS.

Page 19. Fourteenth line from above, after "local name" read—which as a rule is derived either from the English or the Dutch language, except in Vieques and Culebra.

Page 84. To *Avicennia nitida*.—The ground under the tree is sometimes covered with a peculiar kind of aerial roots, proceeding from the underground roots erect into the air to a height of four to six inches.

Page 99. To *Aroideæ*.—A supposed Aroidea with an immense, nearly aphyllous, climbing, terete, green stem, about 100' long, 1" diam., with scaly, early deciduous leaves and aerial roots resembling those of Vanilla, is met with in a few places in St. Thomas (among rocks on Flaghill in the forest). As, however, neither fruit nor flower has as yet been found, it is still doubtful even to which family this interesting species may belong.

Page 100, No. 827. Cancel the lines, "Leaves used for making ropes, thatching roofs, and other domestic purposes."

Add before No. 828 :

827ᵃ. *Th. parviflora*, Sw. (v. Bull-Seger). Fl. May–July; stem 30'–40' high, up to 3' in circumference. Berry in both species black, fleshy. Leaves of this species are used for making ropes, hats, roofs, and for other domestic purposes. On the northern slopes of the hills, common.—Virgin Islands. ·

Add before *Commelynaceæ*:
(Cultivated species : *Phœnix spinosa*, Thonning, and *Latania borbonica*, L.)
118

www.ingramcontent.com/pod-product-compliance
Lightning Source LLC
Chambersburg PA
CBHW021822190326
41518CB00007B/701